Also by Michael Guillen

Bridges to Infinity: The Human Side of Mathematics
Can a Smart Person Believe in God?

FIVE EQUATIONS THAT CHANGED THE WORLD

The Power and Poetry of Mathematics

MICHAEL GUILLEN, PH.D.

MJF BOOKS
NEW YORK

Published by MJF Books
Fine Communications
322 Eighth Avenue
New York, NY 10001

Five Equations That Changed the World
Library of Congress Control No. 00-133285
ISBN 1-56731-405-8

This edition published by arrangement with Hyperion.

Designed by Chris Welch.

Manufactured in the United States of America on acid-free paper

MJF Books and the MJF colophon are trademarks of Fine Creative Media, Inc.

QM   12   11   10   9   8   7   6   5

To Laurel,
who changed *my* world for the better

Acknowledgments

For their exceptional talent and tenacity, I wish to thank my researchers, Noe Hinojosa, Jr., Laurel Lucas, Miriam Marcus, and Monya Baker.

For his extraordinary patience, friendship, and wisdom, I thank my literary agent, Nat Sobel. Also, for their enthusiasm, constructive comments, and support, special credit goes to my publisher, Bob Miller, and editor, Brian DeFiore.

For their invaluable assistance, advice, and encouragement, I am indebted to: Barbara Aragon, Thomas Bahr, Randall Barone, Phil Beuth, Graeme Bird, Paul Cornish (British Information Services), Stefania Dragojlovic, Ulla Fringeli (Universitat Basel), Owen Gingerich, Ann Godoff, Heather Heiman, Gerald Holton, Carl Huss, Victor Iosilevich, Nancy Kay, Allen Jon Kinnamon (Cabot Science Library, Harvard University), Gene Krantz, Richard Leibner,

Acknowledgments

Martha Lepore, Barry Lippman, Stacie Marinelli, Martin Matt-müller (Universitatsbibliothek Basel), Robert Millis, Ron New-burgh, Neil Pelletier (American Horticultural Society), Robert Reichblum, Jack Reilly, Diane Reverand, Hans Richner (Swiss Federal Institute of Technology), William Rosen, Janice Shultz (Naval Research Laboratory), John Stachel (Boston University), Rabbi Leonard Troupp, David Vale (Grantham Museum), Spencer Weart (American Institute of Physics), Richard Westfall, L. Pearce Williams, Ken Yanni (Hoover Dam), and Allen Zelon.

If, despite the aid and comfort of these gracious people, I have made any errors, they are entirely my fault, and I thank the vigilant readers who will surely set me straight.

Contents

Mathematical Poetry

*Poetry is simply the most beautiful,
impressive, and widely effective mode of
saying things.*
MATTHEW ARNOLD

Mathematics is a language whose importance I can best explain by starting with a familiar story from the Bible. There was a time, according to the Old Testament, when all the people of the earth spoke in a single tongue. This unified them and facilitated cooperation to such a degree that they undertook a collective project to do the seemingly impossible: They would build a tower in the city of Babel that was so high, they could simply climb their way into heaven.

It was an unpardonable act of hubris, and God was quick to visit his wrath on the blithe sinners. He spared their lives, but not their language: As described in Genesis 11:7, in order to scuttle the blasphemers' enterprise, all God needed to do was "confound their language, that they may not understand one another's speech."

Thousands of years later, we are still babbling. According to linguists, there are about 1,500 different languages spoken in the world today. And while no one would suggest that this multiplicity of tongues is the only reason for there being so little unity in the world, it certainly interferes with there being more cooperation.

Nothing reminds us of that inconvenient reality more so than the United Nations. Back in the early 1940s, when it was first being organized, officials proposed that all diplomats be required to speak a single language, a restriction that would both facilitate negotiations and symbolize global harmony. But member nations objected—each loath to surrender its linguistic identity—so a compromise was struck; United Nations ambassadors are now allowed to speak any one of *five* languages: Mandarin Chinese, English, Russian, Spanish, or French.

Over the years, there have been no fewer than 300 attempts to invent and promulgate a global language, the most famous being made in 1887 by the Polish oculist L. L. Zamenhof. The artificial language he created is called Esperanto, and today it is spoken by more than 100,000 people in twenty-two countries.

However, as measured by the millions of those who speak it fluently and by the historic consequences of their unified efforts, *mathematics* is arguably the most successful global language ever spoken. Though it has not enabled us to build a Tower of Babel, it has made possible achievements that once seemed no less impossible: electricity, airplanes, the nuclear bomb, landing a man on the moon, and understanding the nature of life and death. The discovery of the equations that led ultimately to these earthshaking accomplishments are the subject of this book.

In the language of mathematics, equations are like poetry: They state truths with a unique precision, convey volumes of information in rather brief terms, and often are difficult for the uninitiated to comprehend. And just as conventional poetry helps us to see deep *within* ourselves, mathematical poetry helps us to see far *beyond* ourselves—if not all the way up to heaven, then at least out to the brink of the visible universe.

In attempting to distinguish between prose and poetry, Robert Frost once suggested that a poem, by definition, is a pithy form of expression that can never be accurately translated. The same can be said about mathematics: It is impossible to understand the true meaning of an equation, or to appreciate its beauty, unless it is read in the delightfully quirky language in which it was penned. That is precisely why I have written this book.

This is not so much an offspring of my last book, *Bridges to Infinity: The Human Side of Mathematics,* as it is its evolutionary descendant. I wrote *Bridges* with the intention of giving readers a sense of how mathematicians think and what they think about. I also attempted to describe the language—the numbers, symbols, and logic—that mathematicians use to express themselves. And I did it all without subjecting the reader to a single equation.

It was like sweet-tasting medicine offered to all those who are afflicted with math anxiety, individuals who normally would not have the courage or the curiosity to buy a book on a subject that has consistently frightened them away. In short, *Bridges to Infinity* was a dose of mathematical literacy designed to go down easily.

Now, emboldened by having written a successful book that contains no equations, I have dared to go that one step further. In this book I describe the mathematical origins of certain landmark achievements, equations whose aftereffects have permanently altered our everyday lives.

One might say I am offering the public a stronger dose of numeracy, an opportunity to become comfortably acquainted with five remarkable formulas in their original, undisguised forms. Readers will be able to comprehend *for themselves* the meaning of the equations, and not just settle for an inevitably imperfect nonmathematical translation of them.

Readers of this book also will discover the way in which each equation was derived. Why is that so important? Because, to paraphrase Robert Louis Stevenson: When traveling to some exotic destination, getting there is half the fun.

I hope that the innumerate browser will not be scared off by the zealousness of my effort. Rest assured, though these five equations look abstract, most certainly their *consequences* are not—and neither are the people associated with them: a sickly, love-starved loner; an emotionally abused prodigy from a dysfunctional family; a religious, poverty-stricken illiterate; a soft-spoken widower living in perilous times; and a smart-alecky, high school dropout.

Each story is told in five parts. The Prologue recounts some dramatic incident in the main character's life that helps set the tone for what is to follow. Then come three acts, which I refer to as Veni, Vidi, Vici. These are Latin words for "I came, I saw, I conquered," a statement Caesar reportedly made after vanquishing the Asian king Pharnaces. Veni is where I explain how the main character—the scientist—comes to his mysterious subject; Vidi explains historically how that subject came to appear so enigmatic; Vici explains how the scientist manages to conquer the mystery, resulting in a historic equation. Finally, the Epilogue describes how that equation goes on to reshape our lives forever.

In preparing to write this book, I selected five equations from among dozens of serious contenders, solely for the degree to which they ultimately changed our world. Now, however, I see that the stories attached to them combine fortuitously to give the reader a rather seamless chronicle of science and society from the seventeenth century to the present.

As it turns out, that is a crucial period in history. Scientifically, it ranges from the beginning of the so-called Scientific Revolution, through the Ages of Reason, Enlightenment, Ideology, and Analysis, during which science demystified each one of the five ancient elements: Earth, Water, Fire, Air, and Ether.

In that critical period of time, furthermore, we see: God being forever banished from science, science replacing astrology as our principal way of predicting the future, science becoming a paying profession, and science grappling with the ultra-mysterious issues of life and death and of space and time.

In these five stories, from the time when an introspective young Isaac Newton sits serenely beneath a fruit tree to when an inquisitive young Albert Einstein nearly kills himself scaling the Swiss Alps, we see science wending its way from the famous apple to the infamous A-bomb. Which is to say, we see science going from being a source of light and hope to its also becoming a source of darkness and dread.

Writers before me have chronicled the lives of some of these five scientists—all too often in frightfully long biographies. And writers before me have reconstructed the pedigree of some of these intellectual innovations back to the beginning of recorded history. But they have never focused their roving attentions on the small number of mathematical equations that have influenced our existence in such profound and intimate ways.

The exception is Albert Einstein's famous energy equation $E = m \times c^2$, which many people already know is somehow responsible for the nuclear bomb. But for all its notoriety, even this nefarious little equation remains in the minds of most people scarcely more than a mysterious icon, as familiar yet inexplicable as Procter & Gamble's corporate logo.

What exactly do the letters E, m, and c stand for? Why is the c squared? And what does it mean for the E to be equated with the $m \times c^2$? The reader will learn the surprising answers in "Curiosity Killed the Lights."

The other chapters deal with scientists less well known than Einstein but who are no less important to the history of our civilization. "Between a Rock and a Hard Life," for example, concerns the Swiss physicist Daniel Bernoulli and his hydrodynamic equation $P + \rho \times \frac{1}{2} v^2 = \text{CONSTANT}$, which led ultimately to the modern airplane. "Class Act" is about the British chemist Michael Faraday and his electromagnetic equation $\nabla \times E = -\partial B/\partial t$, which ultimately led to electricity.

"Apples and Oranges" tells the story of the British natural philosopher Isaac Newton and his gravitational equation $F = G \times M \times m \div d^2$—which led not to any specific invention but to an epic event: landing a man on the moon.

Finally, "An Unprofitable Experience" is about the German mathematical physicist Rudolf Julius Emmanuel Clausius and his thermodynamic equation (or more accurately, his thermodynamic *inequality*) $\Delta S_{universe} > 0$. It led neither to a historic invention or event but to a startling realization: Contrary to popular belief, being alive is unnatural; in fact, all life exists in defiance of, not in conformity with, the most fundamental law of the universe.

In my last book, *Bridges to Infinity*, I suggested that the human imagination was actually a sixth sense used to comprehend truths that have always existed. Like stars in the firmament, these verities are out there somewhere just waiting for our extrasensory imagination to spot them. Furthermore, I proposed that the *mathematical* imagination was especially prescient at discerning these incorporeal truths, and I cited numerous examples as evidence.

In this book, too, readers will see dramatic corroboration for the theory that mathematics is an exceptionally super-sensitive seeing-eye dog. Otherwise, how can we begin to account for the unerring prowess and tenacity with which these five mathematicians are able to pick up the scent, as it were, and zero in on their respective equations?

While the equations represent the discernment of eternal and universal truths, however, the manner in which they are written is strictly, provincially human. That is what makes them so much like poems, wonderfully artful attempts to make infinite realities comprehensible to finite beings.

The scientists in this book, therefore, are not merely intellectual explorers; they are extraordinary artists who have mastered the extensive vocabulary and complex grammar of the mathematical language. They are the Whitmans, Shakespeares, and Shelleys of the quantitative world. And their legacy is five of the greatest poems ever inspired by the human imagination.

$$F = G \times M \times m \div d^2$$

Apples and Oranges

Isaac Newton and the Universal Law of Gravity

I sometimes wish that God
were back
In this dark world and wide;
For though some virtues he might
lack,
He had his pleasant side.
—GAMALIEL BRADFORD

For the last several months, thirteen-year-old Isaac Newton had been watching with curiosity while workmen built a windmill just outside the town of Grantham. The construction project was very exciting, because although they had been invented centuries ago, windmills were still a novelty in this rural part of England.

Each day after school, young Newton would run to the river and seat himself, documenting in extraordinary detail the shape, location, and function of every single piece of that windmill. He then would rush to his room at Mr. Clarke's house to construct miniature replicas of the parts he had just watched being assembled.

As Grantham's huge, multiarmed contraption had taken shape, therefore, so had Newton's wonderfully precise imitation of it. All that remained now was for the curious young man to come up with something, or someone, to play the role of miller.

Last night an idea had come to him that he considered brilliant: His pet mouse would be perfect for the part. But how would he train it to do the job, to engage and disengage the miniature mill wheel on command? That was what he had to puzzle out this morning on his way to school.

As he walked along slowly, his brain raced toward a solution. Suddenly, however, he felt a sharp pain in his gut; his thoughts came to a screeching halt. As his mind's eye refocused, young Newton came out of his daydream and beheld his worst nightmare: Arthur Storer, the sneering, taunting school bully, had just kicked him in the stomach.

Storer, one of Mr. Clarke's stepsons, loved to pick on Newton, teasing him mercilessly for his unusual behavior and for fraternizing with Storer's sister, Katherine. Newton was a quiet and self-absorbed youngster, generally preferring the company of his thoughts to that of people. But whenever he did socialize, it was with girls; they were tickled by the doll furniture and other toys he made for them using his customized kit of miniature saws, hatchets, and hammers.

While it was common for Storer to call Newton a sissy, on this particular morning, he was insulting him for being so *stupid*. Unfortunately, it was true that Newton was the next-to-lowest ranking student in the whole of Grantham's Free Grammar School of King Edward VI, seeded well below Storer. But the idea of this big bully thinking of himself as intellectually superior made the reclusive young man's thoughts turn from windmills to revenge.

As he sat at the back of the class, Newton usually found it easy to ignore what his teacher, Mr. Stokes, was saying. This time, however, he listened with interest. The universe was divided into two realms, each obeying a different set of scientific laws, Stokes instructed. The imperfect, earthly region behaved one way, and the perfect, heavenly region behaved another; both domains, he added, had been successfully studied and their respective ordinances deduced long, long ago by the Greek philosopher Aristotle.

For young Newton, suffering at the hands of an earthly imperfection such as Storer was proof enough of what Mr. Stokes was talking about. Newton hated Storer and his classmates for not liking him. Above all, he hated himself for being so unlikable that even his own mother had abandoned him.

God was the only friend he had, the pious young man thought, and the only friend he needed. Newton was a much smaller person than Storer, but with God's help, he certainly would be able to vanquish the offensive tormentor.

No sooner had Mr. Stokes dismissed class that day than Newton was out the door, waiting in the nearby churchyard for the bully. Within minutes, a boisterous crowd of students gathered round. Stokes's son selected himself referee, slapping Newton on the back as if to encourage him, while winking at Storer as if to say this was going to be as entertaining as watching Daniel being fed to the lions.

At first, no one cheered for young Newton. Instead, each time Storer landed a punch, the rowdy students whooped it up, egging on the ruffian to hit even harder the next time. When it seemed as if Newton had been beaten into submission, Storer straightened up and relaxed, grinning boastfully at his young peers.

As he turned to walk away, however, Newton struggled to his feet: He was not about to let Storer win the right to lord over him for the rest of his life. Alerted by shouts of warning, Storer wheeled around and was greeted with a kick to his stomach and a punch to the nose; Newton had drawn blood, and that reinvigorated him.

For the next several minutes, the two traded blows and wrestled one another to the ground. Time and again, Storer staggered away, thinking he had defeated Newton, only to be confronted anew.

When it was all over, the crowd was stunned into silence. As the young referee stepped in to congratulate the bloodied and exhausted Newton, however, the dumbstruck students stirred and began to cheer: Daniel had become David, they declared jubilantly, as they danced around the fallen Goliath.

Newton was more than satisfied with what he had done, but his schoolmates were not. As he attempted to walk away, young Stokes grabbed his shoulder and encouraged him to humiliate Storer. Newton hesitated, but wishing to gain the approval of his fellow students, he dragged the bewildered bully by the ears and slammed his face into the church wall. The crowd of young spectators squealed with delight as they swarmed around the dazed victor, patting him on the back and accompanying him all the way home with unrestrained shouts of celebration.

Having defeated Storer, Newton's attention quickly returned to the problem of training his pet mouse. Unfortunately for Newton, though, this meant returning to the behavior that had incited his tormentor in the first place.

In a matter of weeks, the still-bruised and -battered Storer worked up enough courage to begin reprising some of his old gibes. Worst of all, Storer's accusations still hit home: Despite his pugilistic victory, Newton remained the dunce of his class.

All his life, with God's help, young Newton had been able to withstand the hazing from insensitive oafs like Storer. But now that he had known the pleasure of being accepted by his compeers, of being loved, he found Storer's effrontery unbearable. This time, he would truly finish the job he had only started in the churchyard.

In the months ahead, Newton paid attention in class as never before and studied his lessons at home. He submitted his completed homework on time and answered all of Mr. Stokes's schoolroom queries.

Gradually, miraculously, one desk at a time, young Newton earned his way to the head of the class. Literally now, he smirked inwardly, he could turn his back on everyone who had ever hurt his feelings or dared suggest they were better or smarter than he.

In the decades ahead, the scope of Newton's interests would expand from windmills to the universe as a whole. But one thing about him would never change: He would meet other antagonists—or people whom he perceived as antagonists—and each time, his obsessive desire for revenge and approval would impel him to an unprecedented understanding of the natural world.

Above all would be his unprecedented understanding of gravity, the force that had always kept our feet on the ground. Newton's stunning disclosure would *sweep* us off our feet, and in the end, our cherished notions about God and Heaven would come toppling down, just like the bully Storer.

VENI

Hanna Ayscough Newton was beside herself with anxiety. Her husband, Isaac, had left suddenly to rally round King Charles I, who had been driven out of London by riotous mobs and an angry, power-hungry Parliament. The king had sought refuge in Nottingham, only thirty miles away from the Newtons' hometown of Woolsthorpe, and from there had just declared war.

England had been involved in many hostilities, but none like this one. This was a declaration of civil war, pitting family members against one another. Ostensibly the conflict was over who would govern England—the royal sovereign or Parliament—but more fundamentally, it was a showdown between heaven and earth.

For centuries, monarchs the world over had been anointed by their country's highest-ranking religious figure; in England, it was the Archbishop of Canterbury. This was no mere ceremony; it was an acknowledgment that kings and queens were selected for office by God himself.

In politics, as in science, therefore, much of the seventeenth-century world consisted of two dramatically separate realms. Mere mortals inhabited the *earthly* realm, but kings and queens were above it all; they dwelt in some lofty, *heavenly* domain, exempt from obeying the strict rules and regulations they imposed on their subjects—and their parliaments.

Over the years, these heavenly appointed rulers had tussled with their earthly appointed parliaments about the details of day-to-day political power. In this regard, Charles had been no different; but now in the fall of 1642, for the first time ever, the two realms had gone to war over the issue of who was preeminent.

Parliament was demanding that Charles relinquish his control over church and state, faulting him for having levied taxes illegally and for having been so religiously intolerant that Pilgrims were now fleeing en masse to uncivilized colonies in America. "The question in dispute between the King's party and us," the rebellious Parliamentarians declared, "was whether the King should govern as a god by his will . . . or whether the people should be governed by laws made by themselves."

In response to this mutinous uprising, Charles had fled from his castle; in Nottingham, he had organized an army of loyalists and was now advancing toward London. Though he and his army were well equipped and fired up, however, their first major battle against the parliamentary forces ended in a draw and left 5,000 soldiers dead.

Among them was thirty-six-year-old Isaac Newton, a yeoman farmer whose father had prospered under the king's controversial yet largely peaceful reign. It had been only last year that Newton had inherited his father's sizable manor—the largest in Woolsthorpe—and only this spring that he had married Hanna and conceived their first child.

Hanna was six months' pregnant when she received the devastating news. She understood and respected the importance of the king's war with Parliament, but she was alternately angry and grief-stricken that her husband had gotten himself killed and orphaned their child-to-be.

The only thing that consoled her was the common belief among villagers that posthumous children invariably grew up having special curative powers and particularly good fortune. She was even more heartened when she gave birth on December 25; a posthumous child born on Christmas Day, the villagers exclaimed, was destined most certainly to be someone very, very special.

No sooner had she laid eyes on the newborn, whom she named Isaac, however, than Hanna began to worry that the locals' joyous predictions would prove to have been premature. Her baby had been born several weeks too early; he was no bigger than a quart jar and gave every indication that he would not survive.

As the pessimistic news spread, the good folks of Woolsthorpe began to speak in hushed tones of a good omen gone bad. Two women sent on an errand on behalf of the newborn, in fact, did not bother to walk very quickly and rested many times along the way, so certain were they that the ill-fated child would die before they returned.

They were wrong. As the days passed, baby Isaac clung to life with increasing strength, revealing a stubbornness, a willpower so extraordinary the villagers appeared to have been vindicated after all: This son of a dead man, born on Christ's birthday, they whispered, was no ordinary human being.

During the first few years of his life, young Isaac Newton was so feeble he had to wear a neck brace to hold his head in place. Nevertheless, the danger to his life had passed, and everyone in Woolsthorpe assumed that mother and child would settle into a reasonably happy and comfortable existence.

Once again they were wrong. When Newton was only two years old, his mother received a proposal of marriage from the Reverend Barnabas Smith, a wealthy, sixty-three-year-old widower from North Witham, a town located about a mile away. After consulting her brother, the Reverend William Ayscough, Hanna decided to accept, whereupon she moved to North Witham *without* her son, whom she left in the care of her mother.

Being abandoned at such a young age would have been traumatic enough under normal circumstances. But this was 1645, and England's civil war was now raging throughout the countryside. Woolsthorpe, at first under the protection of the king, had been captured by Parliament. Every week, there was gunfire from mortal skirmishes being fought in the area and intrusions from raiding parties in search of provisions and billeting. All this chaos frightened the frail young Newton, and worse, when he cried for his mother, she wasn't there to comfort him.

Newton's grandmother tried her best to mollify him, but she herself was quite frightened by what was happening. Nearly all the able-bodied men of Woolsthorpe had been killed or called away to fight, leaving only clergymen to help defend the women and children against the bestialities of the warring armies.

Adding to his fright, in 1649, the youngster began to attend school. Being delicate by birth, he was afraid (and not welcome) to participate in the aggressive games played by the other boys. Being an orphan, moreover, he felt inferior to the other children, most of whom lived in homes enriched by the love of a mother and father.

He was even more discomfited later that same year when the village received news that the Puritan-dominated Parliament, led by Oliver Cromwell, had defeated the royal armies; King Charles himself had been beheaded. Over the years, young Newton had formed a vicarious attachment to the swashbuckling monarch, fully expecting that one day this surrogate father figure would come galloping in to rescue him and his village from the nasty Parliamentarians.

It was during these perilous years that young Newton came to cherish the companionship of his uncle, Hanna's brother, who lived two miles away. Like all Anglicans at the time, the Reverend Ayscough saw the civil war in religious terms, pitting the king—England's "Defender of the Faith"—against a Parliament controlled by Puritans.

Both sides were devoted Christians, of course, but they were split as to the way in which organized religion should be governed. Anglicans were administered by a hierarchy of clergymen, headed by the Archbishop of Canterbury, the English equivalent of the Pope. The Puritans were organized in a less hierarchical, more purely democratic, fashion. In truth, their differences were rather esoteric, yet mutual intolerance was causing them to kill one another.

Newton was far too young to understand any of this, but as he watched his uncle studying peacefully in the library, listened to his uncle speaking gently to his parishioners, young Newton became conditioned to associate a religious and scholarly lifestyle with safety and security.

In a short time, therefore, young Newton acquired the habit of turning away from the encircling chaos and toward his own thoughts. He sought out secluded areas, where he would sit for hours at a time, not so much to observe the natural world as to *immerse* himself in it.

The young man discovered that if he meditated single-mindedly on the minutiae of his surroundings, he was able to escape from his miserable existence and discover interesting things about Nature. For example, he noticed, rainbows always came in the same colors, Venus always moved faster than Jupiter across the night sky, and children playing ring-a-ring o'roses invariably leaned a bit backward, as if they were being nudged by some invisible force.

In these wholly encompassing immersions, the youngster was able to enter a sanctuary every bit as comforting as his uncle's rectory, without having to travel the two miles to get there. Best of all, he discovered true happiness for the first time in his life.

In 1649, Newton's newfound rapture was spoiled by the return of his mother and several young strangers. The Reverend Barnabas had died, but only after having fathered three small children, one of them less than a year old. Even now, even with his mother's return, young Newton fumed and fussed, he would not have her undivided love and attention.

During the first several months of her homecoming, Mrs. Newton-Smith tried to explain to her angry son that she had married the old rector solely to secure their long-term financial security. The rector of North Witham, she revealed, had paid for a renovation and expansion of the Newton's manor and bequeathed to young Newton a large parcel of land.

Nothing his mother said, however, could sweeten his bitterness at having been abandoned. Newton hated his mother and often had dreamed of setting fire to her and her second husband while they lay sleeping together.

For the next few years, therefore, though one civil war between king and Parliament had ceased, another raged between mother and son. Ultimately, the only thing that stopped it was forcible separation: This time, however, it was young Newton who left his mother.

It had come time for the twelve-year-old to attend grammar school in the city of Grantham, seven miles away. Since that was too far to walk, his mother had arranged for him to room and board with the Clarke family, longtime friends of the Newtons.

Having lived with a mother he hardly knew and three half siblings he didn't care to know, young Newton was unfazed by the idea of moving in with complete strangers; at least, he thought, they gave the appearance of being an honest-to-goodness family. There was Mr. Clarke, who ran his own apothecary; Mrs. Storer-Clarke and her four children from a previous marriage; they included a pugnacious son named Arthur and an attractive daughter, Katherine, who took an instant liking to the new boarder.

The Clarkes frequently entertained learned guests, so Newton's mind was kept well fed with food for thought. Most wonderful of all was Mr. Clarke's vast collection of books in the attic. Here was the perfect getaway, the ideal sanctuary, Newton enthused, as he proceeded to immerse himself in subjects ranging over the entire intellectual spectrum.

The books and dinner guests had the salutary effect of introducing this lonely youngster to a world of kindred spirits: the Frenchman René Descartes, who offered a theory for the recurring colors of the rainbow; the German Johannes Kepler, who discovered that a planet moved more slowly the farther away it was from the sun; and the Dutchman Christiaan Huygens, who gave the name *centrifugal force* to the ring-a-ring o'roses phenomenon young Newton had noticed a few years earlier.

Just that suddenly, Newton had the inklings of what it was like to feel normal. All his life, he had felt like an intruder, as if there were no place for him on this earth. Now, in the study of natural philosophy, he had found a home, a community of persons like himself, where he might be accepted, appreciated, possibly even loved.

During this time, Newton fell behind in his studies at school, so distracted was he by his newly adopted intellectual family. Also, it didn't help his concentration that he had become infatuated with Mr. Clarke's comely and kind stepdaughter Katherine—though he was too shy to express his feelings except through the toy furniture he made for her.

Indeed, it took a kick in the stomach from the girl's bully of a brother to awaken young Newton from his reverie and to coax him into scrapping his way to the head of the class. However, no sooner had he done so than his mother intruded once again; this time, she ordered him back to the manor.

The properties and responsibilities the Newtons had inherited from the late Reverend Smith had become too burdensome for her to manage alone. Besides, she remonstrated, her son had already received more than an adequate education; after all, neither his father nor any other Newton in history had even been able to write their own names.

Newton returned to Woolsthorpe, but he did so over the protests of his teacher and uncle. Not only was Newton now the school's top student, Stokes and the Reverend Ayscough pleaded, but by having earned that distinction so dramatically, the young man was quite possibly the first bona fide genius this rural county had ever produced.

The teenager now disliked his mother more than ever; he was openly disobedient and frightfully surly. As a symbol of his protest, the seventeen-year-old Newton purchased a small notebook: His body might be back in Woolsthorpe, he thought defiantly, but his mind would remain on natural philosophy, which required all its students to keep a careful journal of their theories and observations.

Unfortunately for Hanna Newton-Smith, but fortunately for science, her son proved to be inept at running a farm. One day, for example, he became so engrossed in a small water wheel he had built, he did not notice that a group of pigs had forded the stream and were eating the neighbor's corn.

His mother was fined "for suffering his swine to trespass in ye corn fields," the court clerk wrote in the record, and "for suffering his fence belonging to his yards to be out of repair." That was not the first time Mrs. Newton-Smith had had to pay for her son's distractedness, but it was most certainly to be the last; forthwith, she sent him packing back to Grantham.

No sooner had young Newton returned to the Clarke household than he realized fully just how much he had missed not only his studies but the lovely Katherine. She herself gave many an indication of having similar feelings toward him—a gentle touch here, a kindly glance there—but all to no avail. So fearful was he about being rejected, Newton stopped short of ever confessing his romantic feelings to her.

The young man was much more aggressive when it came to his grammar schooling, finishing it in only nine months. On his last day, in the summer of 1661, Mr. Stokes bid him to stand before the class. As the young man obeyed, he and his classmates had the impression that a scolding was about to take place. There were furtive glances, whispers, and a lot of fidgeting. But why? What now! Newton wondered glumly.

Facing the class, expecting the worst, Newton was soon relieved of his anxiety. Mr. Stokes began praising him for being such a model student, entreating the others to be like this young man who, though orphaned, bullied, and badgered, had become the pride and joy of Lincolnshire County. Weeping, the devoted teacher delivered such a moving tribute to his prize pupil that even the young students seated at their desks had tears in their eyes when it was all over.

On the strength of enthusiastic recommendations from the Reverend Ayscough and Mr. Stokes, not to mention the merits of his own achievements, young Newton was readily accepted into Trinity College, the reverend's alma mater. It was, as he put it in a letter to his mother, "the famousest College" on the entire campus of Cambridge University, having been founded in 1546 by none other than King Henry VIII.

Objectively speaking, seventeenth-century Cambridge was little more than a dingy village, but to this young man from the country, it was the grandest place he had ever seen. By coincidence, it was also at its gayest in more than a decade.

Eleven years earlier, when the civil war had been decided in favor of Parliament, the puritanical victors had imposed on England unprecedentedly strict rules of behavior. They had made adultery a capital crime and outlawed nearly all manner of recreation, including horse racing, theater, and dancing round the Maypole. The Puritan rulers even had outlawed the celebration of Christmas, prompting one aghast Anglican to grouse: "Who would have thought to have seen in England the churches shut and the shops open on Christmas Day?"

By 1660, the English had had enough of being forced to live so austere an existence—of obeying the severe rules of some puritanical heavenly realm, as it were. They yearned for the more frolicsome rules of the delightfully imperfect earthly realm, whereupon they restored the sacred English crown to Charles II, the beheaded king's eldest son. Thus, in 1661, when Newton arrived in Cambridge, he found it in the midst of celebrating the country's return to a more secular existence, complete with parades, music, and rowdy fairs.

While England was loosening its hair, though, young Newton was obliged to tighten his belt. Mrs. Newton-Smith was more than wealthy enough to pay for her son's tuition, but she had decided to withhold her support, forcing the freshman to be enrolled into the college as a subsizar.

This was the name given to poor students who helped finance their education by being part-time servants to others whose parents fully supported them. For the next several years, therefore, Newton once again found himself being tormented by equals who felt superior to him; moreover, it would have been easier to withstand the abuse if, deep down inside, Newton himself had not felt inferior and unloved.

Instinctively the young man reverted to his old habits. Whenever he was not occupied with classes, church services, or his servile duties—which included emptying chamber pots, grooming his master's hair, and hauling firewood—the insecure prodigy from Woolsthorpe immersed himself in the details of the natural world.

One evening, after finishing his subsizar's chores in the kitchen at Trinity, he divided the heart of an eel into three sections. For hours, the young man stared and took careful notes, marveling at how the disconnected pieces continued to beat in synchrony.

Newton even began to experiment on his own eyes with harrowing carelessness. At one point, he wedged a flat stick "betwixt my eye & ye bone as neare to ye backside of my eye as I could," coming dangerously close to blinding himself, all in the hopes of understanding exactly how humans perceive light and color. "Pressing my eye with the end of it . . . there appeared several white, darke & colored circles," he noted casually, "which circles were plainest when I continued to rub my eye with ye point of ye bodkin."

During his years at Trinity, his small notebooks, which he carried with him everywhere, came to be filled with the observations and queries of his powerful concentration and wide-ranging curiosity. "Of Light and Color," "Of Gravity," "Of God"—these were more than mere headings in this queer young man's present investigations, they were glimpses at the voracious appetite of a rare and gifted mind.

While Newton's brain sped onward, well nourished and full of energy, his body began to lag behind; in 1664, it gave out altogether. His ceaseless inquiries having deprived him of sleep for the better part of his undergraduate career, Newton was bedridden with exhaustion.

Though he felt weak for many months thereafter, the young man recovered in time to take his final exams. He did not perform well, but he earned his bachelor of arts degree. Moreover, influential professors who espied in this introverted and mediocre student the makings of a first-rate scholar intervened, and Newton was granted a scholarship to pursue a master's degree.

He had hardly commenced his new course of studies when news reached Cambridge that the dreaded plague had invaded London. In the past twenty years, that city's population had doubled, seriously compromising its medieval sanitary facilities. Now reports indicated that up to 13,000 people a week were dying.

Though Cambridge was more than forty miles away from all of that, officials decided anyway to close down the university, wishing to avoid a repetition of history: Back in the fourteenth century, the Black Death, as it was called, had spread like a pestilent wind all across Europe, turning Cambridge into a ghost town.

Before the formal order was given for students to evacuate the city, however, young Newton had returned to Woolsthorpe: Even his mother's company was preferable to the risk of being killed by this horrific affliction. Anyway, he figured, it was time to reflect on all that he had learned during the past four years at Trinity.

It was the summer of 1665, and while hysteria and death roiled through the narrow streets of London, the twenty-two-year-old spent his days lounging in the garden, puzzling out the details of a new mathematics that would one day come to be called the *calculus*. Above all, he savored the solitude, his mother having long since given up on nagging him to become a gentleman farmer.

On one particular day, the weather was so agreeable and Newton was so immersed in thought, he did not notice it was getting late. Gradually the garden around him began to glow warmly, bathed with the soft golden light only a waning summer sun could produce.

Suddenly the thud of an apple falling from a nearby tree startled the young man out of his deep meditations. In the few moments it took for him to switch trains of thought, the top of a gigantic-looking full moon began to show itself above the eastern horizon.

Within minutes, young Newton's insatiable curiosity began to nibble away at the apple and the moon. Why did apples fall straight down to the earth's surface, rather than askance? What if the apple had started from higher up—a mile, a hundred miles, as high as the moon—would it still have fallen to the earth?

For that matter, didn't the moon itself feel the tug of earth's gravity? If so, would it not mean that the moon was under the sway of earthly influences, which contradicted the common belief that the moon existed within the heavenly realm, completely aloof from our planet?

Engaged by these heretical speculations, Newton persevered into the wee hours of the night. If the moon could feel the earth's tug, then why didn't it fall to the ground like an apple? No doubt, he conjectured, it was because Huygen's centrifugal force pulled the moon *away* from the earth; if that and the earth's pull balanced each other out, then perhaps that would explain how the moon was able to stay in its ring-a-ring o'roses orbit indefinitely.

Seated beneath the steely light of the moon, Newton was engrossed in his thoughts. More than that, while crickets chirped and frogs croaked in a nearby pond, the young man began to jot down certain ideas and calculations that would one day lead him to formulate his extraordinary equation of universal gravitation.

It would take more than twenty years before the world would learn of what had happened this night. It would take that long for Newton to perfect and publish his results, but when that day came, the heavens would fall to the ground with the thunderous boom of a million plummeting apples.

VIDI

Twenty-three centuries ago, Plato led a historic revolt against the traditional gods who lived atop Mt. Olympus. They were no longer praiseworthy, he complained, because they had become too mischievous, too immoral, and too undignified.

More than that, the famous Academician sniffed, those old gods were now too provincial for a Greek empire that had expanded dramatically under the Macedonian leadership of King Philip II (and soon would grow even larger under his son, Alexander the Great). Such a vast and victorious civilization needed—nay, merited—world-class divinities.

"A man may give what account he pleases of Zeus and Hera and the rest of the traditional pantheon," Plato intoned, but it was time for the Greek people to enlarge their religious horizons by looking heavenward, recognizing the "superior dignity of the visible gods, the heavenly bodies."

As if that were not enough to ask of his fellow countrymen, Plato went on to implore them to "cast off the superstitious fear of prying into the Divine . . . by setting ourselves to get a scientific knowledge of their [i.e., heavenly bodies'] motions and periods. Without this astronomical knowledge," he argued in sublime rhetorical fashion, "a city will never be governed with true statesmanship, and human life will never be truly happy."

Convincing the Greek people to adopt entirely new gods *plus* asserting that mere mortals were capable of comprehending godly behavior was a religious revolution of the most radical sort. It also was a *scientific* revolution, though this was not to be recognized fully until Isaac Newton's dramatic discovery in the seventeenth century.

That recognition was slow in coming, it turned out, because astronomers were slow to interpret correctly what they were seeing in the night sky. The sun, moon, and stars all behaved impeccably, they felt, always appearing to move in perfect circles around the earth; among all known curves, circles were considered godly, because they were flawlessly symmetric and, by virtue of their having no beginning and no end, eternal.

What befuddled astronomers were five spots of non-twinkling light that seemed to wander hither and yon across the night sky as if they were drunk. Plato was aghast: This erratic behavior was not godlike—indeed, it was redolent of Zeus's and Hera's outrageous shenanigans—and it threatened to discredit his religious reformation.

Greek astronomers soon began referring to these wayward deities as *planets*—the Greek word for vagabonds—and set upon trying to make sense out of their seemingly imperfect movements. It took them two decades, but the effort was well spent: Plato's religious revolution was rescued by a heroic exercise in circular reasoning.

Whereas the other heavenly bodies appeared to whiz around in imaginary circles, Plato and his colleagues explained, planets whizzed about with a great deal more freedom upon the surfaces of imaginary *globes*. Since globes were just as symmetric and seamlessly eternal as circles—in fact, mathematically speaking, globes were nothing but two-dimensional circles—planetary motion was no less divine than the motion of the moon, sun, and stars.

In the years following Plato's death in 347 B.C., Aristotle extended his mentor's incipient revolution even further. With extraordinary detail and fabulous logic, Aristotle now offered an explanation for how and why Plato's new celestial gods were superior to humans and everything else on earth.

All the heavenly bodies in the universe—the moon, sun, planets, and stars—revolved around the earth, which itself did not move in any way. Furthermore, Aristotle theorized, the universe was segregated into two distinct regions: The central one encompassed the earth and its atmosphere; beyond that—from the moon outwards—was what Aristotle referred to as the celestial region.

The *earthly* realm, Aristotle opined, consisted of only four essential qualities: wet and dry, hot and cold. They alone underlay everything terrestrial, including the four elements his contemporaries believed were the bases of physical reality. What they called Earth was essentially dry and cold; Water was cold and wet; Air was wet and hot; Fire was hot and dry.

The earthly realm was corruptible and changeable, Aristotle maintained, because the quartet of basic elements and their underlying four qualities were themselves corruptible and changeable. For example, if one heated Water, which was cold and wet, it became Air, which was hot and wet.

Furthermore, Aristotle explained, all four terrestrial elements tended to move in straight lines, which was entirely appropriate: Straight lines were the most earthly of all curves, because they had endpoints, symbolizing birth and death. For example, if not otherwise coerced, Earth and Water always opted to move straight *downward,* giving them an air of gravity. By contrast, Air and Fire appeared to possess an inherent levity, always preferring to move straight *upward.*

The *celestial* realm was another matter altogether. It consisted entirely of a fifth basic element, a quintessential protoplasm named Ether. This miraculous material came in different densities, Aristotle imagined, forming everything from the sun, moon, stars, and planets to a nested set of revolving globes upon whose invisible surfaces the heavenly bodies were whirled around in their flawless orbits.

The moon, sun, and stars were attached to globes that always spun in one direction, which explained their perfectly circular orbits. As for the celestial vagabonds, the planets, they were attached to globes that spun this way and that in an orderly but complicated fashion, which explained their more varied movements across the night sky.

Unlike the four earthly elements, Aristotle believed, Ether was incorruptible. Its flawlessness meant that the heavens would always remain perfect and unchanging; they never would rust or break down.

With this theory of the universe, Aristotle had fulfilled Plato's fondest wishes: He had given the earthly rabble their first peek at the privileged lifestyles of the celestial starlets, fresh-faced deities whose impeccable behavior was at once unassailable *and* comprehensible. People were quite thrilled at what they saw, furthermore, because Aristotle's universe was through and through a *cosmos,* the Greek word for orderliness, beauty, and decency—everything they could have hoped for in their new gods.

His theory also satisfied the Principle of Sufficient Reason, so

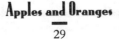

dear to Western philosophy, which maintained that for every effect in the universe, there had to be a rational cause. For example, according to Aristotle, pieces of Earth fell downward because of a natural desire to be reunited with their primary source, the earth. Heavy objects fell faster than light ones, he supposed, because their desire was that much greater.

Aristotle even had a plausible and reverent explanation for what caused the huge heavenly globes to revolve. Each one, he explained, was swept around by an ethereal wind whipped up by the movement of the globe immediately above it, the outermost globe being impelled by the *Primum Mobile,* the prime mover, God Himself.

Plato had introduced religion and science to one another and had lived long enough to see the two engaged. Now Aristotle had married them in a most endearing and enduring way. Moreover, every indication was that this odd couple would benefit mutually from the unprecedented betrothal.

For its part, science painted a flattering picture of the heavens and corroborated the existence of a supreme god. Its down-to-earth explanations of an otherwise mysterious realm informed and enriched people's religious convictions, exactly as Plato had hoped: "The study we require to bring us to true piety," he had said, "is astronomy."

For its part, religion expanded the domain and elevated the reputation of science. Before this, to the extent that it was even definable, science had been widely regarded as an eccentric enterprise of doubtful value, preoccupied by the esoterics of the earthly realm and the abstractions of the mathematical realm.

As the centuries passed, however, so did the Greek empire and the fruits of its historic religious and scientific innovations. The rise of Christianity in the Western world became the newest religious revolution, during which many old earthly gods were exchanged for the one heavenly God worshipped by orthodox Jews and extolled by the recently martyred heretic, Jesus of Nazareth.

Since most people in the civilized world came to speak Latin, not Greek, they lived and died never even knowing about Aristotle, much less his theory of the universe. As the old Greek texts gradually were translated, however, Christians discovered that, as the Dominican St. Albert the Great enthused: "The sublimest wisdom of which the world could boast flourished in Greece. Even as the Jews knew God by the scriptures, so the pagan philosophers knew Him by the natural wisdom of reason, and were debtors to Him for it by their homage."

By the thirteenth century, students across Europe were beginning to learn all about Platonic rhetoric, Aristotelian logic, and Euclidean geometry; indeed, it became quite the fashion to do so. More significantly, Christian leaders were learning that the rabbi Maimonides had already reconciled Aristotle's cosmology with Judaism and that the philosopher Averroës had done the same with the religion of Islam.

Not to be left behind, therefore, the brilliant Dominican theologian St. Thomas Aquinas helped to accommodate Aristotle's geocentric universe to Christianity. There were myriad subtleties involved, but the upshot was that the heavenly bodies, no longer worshipped as demigods, were imagined to ride upon globes kept spinning by angels, not ethereal winds. Above all, Aristotle's *Primum Mobile* was identified with the one-and-only Judeo-Christian God, not just some generic divinity.

What Aristotle had first joined together and time and language differences had put asunder, Jews, Muslims, and now Christians had rejoined. Science and religion were in one another's arms yet again, and this time their honeymoon would last throughout a historic renaissance in Western civilization.

Beginning in the fourteenth century, however, much of the inhabited world was devastated by a succession of horrifying outbreaks of bubonic plague. Between the years 1347 and 1350 alone, it wiped out at least one-third of the European population.

In the aftermath, there was a great deal of finger-pointing: Survivors blamed their spiritual leaders for not having forewarned them of this apocalyptic rebuke from God. And in response, the clergy castigated the masses for inviting such punishment with their sinful behavior.

Ironically, Christian churches and monasteries all over Europe had been hit worse than the general laity; fully half of God's holiest were now dead, which regrettably led to an even greater calamity. As one observer noted: "Men whose wives had died of the pestilence flocked to Holy Orders of whom many were illiterate."

Lured by large sums of money offered by villages bereft of a religious leader, more and more men joined the priesthood for all the wrong reasons. Most of them were "arrogant, given to pomp," Pope Clement VI lashed out in disgust, wasting their ill-gotten wealth "on pimps and swindlers and neglecting the ways of God."

In this derelict and weakened condition, the Catholic church was pummeled by two of its most disillusioned members. In 1517, the German priest Martin Luther fathered a historic religious reformation by beseeching his colleagues to return to a Christianity sustained by childlike faith and good deeds, not by the extravagances of a temporal world. And in 1543, the Polish theologian Nicholas Copernicus touched off a religious-scientific revolution by urging a divorce from Aristotle; at the center of the universe was the sun, he claimed, not the earth.

Copernicus was an amateur astronomer, but he had no new observational evidence with which to defend his opinion. He simply believed that the geocentric theory was unnecessarily complicated, made so by the misguided presumption that we were looking at the heavens from some rock-steady vantage point smack in the middle of the action.

For example, Copernicus speculated, the movement of the vagabond planets appeared complicated only because we ourselves were moving through space in a complicated way, riding aboard an earth that was whirling on its axis like a ballerina as it revolved around the sun. Once we took those earthly motions into account, he demonstrated, the motion of the planets became sublimely circular, just like that of all the other heavenly bodies.

To a child being swung around by the arms, everything in the world appeared to wobble and spin. Were things really moving that way? The child's answer would be "no, of course not" only if he admitted to being the one who was spinning around, not they. Such was Copernicus's simple but stinging argument.

This Polish canon of Frauenburg, East Prussia, was not the first to have championed the heliocentric theory; 2,000 years earlier, a number of Greek philosophers had come up with several versions of the same idea. It had proven to be controversial back then, and for many of the same reasons, it turned out to be so again.

Scientifically speaking, the critics pointed out, it simply did not feel as if the earth were moving; if indeed it was whirling around the sun and spinning around on its axis, we would expect there to be some overt indication of it. Some astronomers even conjectured that everything would be flung off the earth's surface, like water droplets flying off a spinning wet wheel.

Religiously speaking, there were compelling objections as well. In Joshua 10:12–13, the Old Testament clearly stated that during the battle of Gibeon "the sun stood still, and the moon stayed, until the people had avenged themselves upon their enemies." Most who believed in the Judeo-Christian God took this to imply that, quite literally, under normal circumstances, the sun and moon *moved* around the earth.

In view of these and other objections, and because there was no physical evidence to favor the Copernican theory, most of the civilized world—religious and scientific—continued to believe in Aristotle's view of the heavens. Even fellow revolutionary Martin Luther ridiculed Copernicus for defending such an outlandish idea

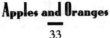

as heliocentricism. Still, it had been a revolutionary century, and before it was over, there appeared signs in the heavens that tended to absolve Copernicus.

The first omen came during one evening in 1572. A bright new star suddenly appeared in the sky (astronomers later believed it to be an exploding star, or supernova), causing people everywhere to look up in wonder. "A miracle indeed," a Danish astronomer named Tycho Brahe gushed, "the greatest of all that has occurred in the whole range of nature since the beginning of the world."

For Aristotle's theory, this miracle was a disaster, because it violated the theory's all-important premise that the heavens were permanent and incorruptible. Only in the earthly realm were things supposed to pass in and out of existence like that.

Five years later, a second omen compounded the disaster. This time it was a comet so bright it could be seen in daylight all over Europe. Amazing as it was, however, the astronomer Brahe was even more stunned when he measured the comet's parallax.

Parallax was an optical illusion that astronomers had found so useful it had become one of the tricks of their trade. When looking at an object, first with the right eye and then the left, an object appeared to shift position with respect to the background. Fortuitously, the amount of that shift, or parallax, *decreased* as the object's distance *increased*. (See it for yourself, by looking at your index finger from different distances.)

In the case of the comet, the right-eye view was provided by Brahe watching from an island off the coast of Denmark. The left-eye view, so to speak, was provided by his colleagues in Prague. The difference between the two views, the parallax, enabled Brahe to conclude that the comet was four times farther from us than the moon.

Astronomers were incredulous. Aristotle had said, and they had always believed, that comets were caused by fiery disturbances in the earth's atmosphere, existing not much farther up than ordinary clouds. For a comet to be streaking through the heavens, beyond the moon, was unthinkable.

On the face of it, therefore, the recent star and comet were unsightly blemishes on Aristotle's stellar reputation. Indeed, the only vindication Aristotle received during these fateful years was in regard to his belief that comets were harbingers of doom. On that score, unfortunately for his theory of the universe, he was absolutely right.

In the years ahead, as science became increasingly receptive to the possibility that Aristotle might be wrong, religion became more defensive toward dissenters of any kind. Luther's religious reformation had spawned a vast Protestant movement, and the orthodoxy felt threatened and reacted like a wounded animal with its back against the wall.

In 1600, the Italian monk Giordano Bruno was burned at the stake for his belief in a bizarre philosophy that was part Christianity and part alchemy, among other things. He also happened to believe in the Copernican theory of the universe, and because of that coincidence, his grisly execution had a chilling effect even on those religiously pious individuals who questioned the earth's central position but not the church's central authority.

Especially upset were Christian scientists who believed their religion could be reconciled with Copernicus's heliocentricism, just as once before it had been reconciled with Aristotle's geocentricism. Now, however, they were reluctant to express their opinion openly, for fear of attracting the attention of inquisitors whom the Catholic church recently had empowered to prosecute heretics.

A forty-seven-year-old German astronomer named Johannes Kepler was being particularly discreet, because he was a Lutheran *and* a Copernican—the worst of all worlds, so far as the religious establishment was concerned. He was director of the observatory originally run by the late Tycho Brahe, furthermore, and was about to announce several discoveries that were sure to finish up where Brahe's assaults had left off in discrediting Aristotle's theory of the universe.

Getting that far in life had not been easy for Kepler. He was only sixteen when his father had deserted the family, relegating them to a life of abject poverty. To make matters worse, his mother had been rumored to be a witch, and that ultimately had cast malevolent suspicion on Kepler and his predilection for astrology.

Kepler was indeed a remarkable astrologer; during one year he correctly predicted a cold winter, a peasant uprising, and a Turk invasion. As a scientist, however, he tended to downplay his success: "If astrologers sometimes do tell the truth," he had explained demurely, "it ought to be attributed to luck."

Kepler preferred astronomy, but there were no jobs for someone who wished to study the stars solely for their scientific value. Therefore, as a young man trying to earn a living for himself and his poor, bedeviled mother, he had found it lucrative to cast horoscopes. Besides, Kepler had always harbored the vague and generic belief that heavenly bodies do somehow influence earthly affairs.

For example, at the age of six, he and his mother had stood outside, gazing fearfully and excitedly at the glowering comet of 1577. He hadn't thought of it again until years later, when Brahe, the great comet observer himself, had offered the impecunious young astronomer a job. From that moment on, Kepler had never lost the intuitive conviction that his present position of prominence had been foreshadowed by that comet.

This now was his finest hour. He had spent the better part of the past twenty years trying to make sense of Tycho Brahe's meticulous observations of the heavens. Using the very latest equipment (save for a telescope, which had yet to be invented), Kepler himself had spent hundreds of hours observing the planets, trying to discern their true "periods and motions," just as Plato had once enjoined his countrymen to do.

Now, two millennia later, the mission had been fulfilled, but the outcome was nothing like what Plato or Aristotle had foreseen. Kepler had discovered three remarkable things about the vagabond planets, the first being that they do indeed have a wonderful underlying regularity to their behavior, *if* one believed that the sun sat at the center of them all.

If *T* stood for the length of a planet's year (the time it took for a planet to go once around its orbit) and *d* stood for the planet's distance from the sun, then the first thing Kepler had discovered boiled down to this simple equation:

$$T^2 = \text{constant} \times d^3$$

In plain English: The square of a planet's year always equaled some multiple of the cube of the planet's distance from the sun. That is, planets far from the sun had long years, whereas planets close to it had short years. (Mercury, the planet closest to the sun, has a year 88 days long; Pluto, farthest away, has a year 90,410 days long!)

The second thing Kepler had discovered revealed an unseemly irregularity in the heavens. Planets did not move at constant speeds along their orbits, he announced; instead, they routinely sped up and slowed down, like a jockey who kept changing his mind about how fast to race around the track.

Finally, Kepler disclosed, the planets raced around orbits that were shaped like *ovals,* not circles! Of the three revelations, this one pierced closest to the very heart of Aristotle's age–old vision of heavenly perfection.

Though these discoveries were dangerously sacrilegious, at this particular moment, the middle-aged astronomer did not care: "Now . . . nothing holds me back. I yield freely to the sacred frenzy. If you pardon me, I shall rejoice," a deliriously happy Kepler enthused with reckless abandon, "if you reproach me, I shall endure."

In the years ahead, Kepler was able to avoid the Catholic Inquisitors and to concentrate on honing his defense of heliocentricism. In his opinion, for instance, planets were kept in their orbits, not by ethereal globes, but by some kind of magnetic force from the sun.

His contemporaries had different theories: The French philosopher René Descartes, for example, believed that all heavenly bodies were located at the tapered ends of giant, invisible tornadoes. The planets spun around the sun, he supposed, simply because they were caught up in the sun's whirlwind.

By the same token, Descartes explained, the moon spun around the earth because it was caught up in the earth's invisible tornado. Furthermore, things *fell* to earth whenever they were unfortunate enough actually to be *sucked in* by the tornado.

In Florence, Italy, yet another astronomer, a sixty-nine-year-old named Galileo Galilei, was getting caught up in the swirling winds of change. Like Kepler and nearly everyone else in his generation, Galileo had begun life as an avowed Aristotelian. But he had changed his mind back in 1609, when he had looked through a crude little telescope of his own design; with it, he had beheld tiny moons circling around Jupiter, exactly as Copernicus had imagined the moon circling around the earth.

The earth's moon, furthermore, was not nearly as perfect as Aristotle had pictured; it was full of large blemishes. Some looked like craters, Galileo commented, and others looked like *maria,* seas full of *water,* a corruptible element supposedly found only within the earthly realm.

(Many years after Galileo had been proven wrong about the water, scientists would retain his imagery. In fact, the first lunar astronauts would land within an area called *Mare Tranquilitatis,* or the Sea of Tranquillity.)

Galileo also had found powerful reasons right here on earth to doubt Aristotle. For example, in measuring how fast metal balls rolled down slopes of various steepnesses, he had discovered that heavy objects did *not* fall faster than light ones; contrary to common sense, and to Aristotle's vaunted theory, all objects fell to the earth with identical speeds.

Unfortunately for Galileo, living in the national seat of Roman Catholic power meant that he was in greater danger than Kepler of being accused of heresy for his often tactless denunciations of Aristotle and brash belief in Copernicus's heliocentricism. It was not surprising, therefore, when in 1633 he was summoned to the Vatican to face the Inquisition.

Galileo stood accused of ignoring an order from the Holy Office issued to him fifteen years earlier. The papal injunction had warned "that the said opinion of Nicolaus Copernicus was erroneous" and that, therefore "the aforementioned Galileo was ordered and enjoined to abandon completely this opinion. He is not to hold, to teach or to defend it in any way whatever, either orally or in writing."

Though he insisted throughout the months-long trial that his belief in heliocentricism had always been purely academic, Galileo could not deny having defied the letter and spirit of the church's ominous warning. Consequently, on June 21, 1633, a tribunal of cardinals found him guilty and demanded that he recant.

At first, Galileo was intransigent. "I have nothing to say," he insisted. But after being threatened with the same fate as Giordano Bruno, the weary old astronomer relented: "I, Galileo, being in my seventieth year, being a prisoner and on my knees, and before your Eminence, having before my eyes the Holy Gospel, which I touch with my hands, abjure, curse, and detest the error and the heresy of the movement of the earth."

Badgered further, the defeated and prostrate Galileo reiterated his confession: "I do not hold this opinion of Copernicus," he said with a moan. "For the rest I am in your hands. Do with me as you please."

The tensions that had begun to strain the long-standing marriage between science and religion had finally erupted into a very ugly public brawl. It was, however, not what it seemed: In Rome, religion had brought science to its knees, but in reality, it was science that now threatened to fell religion.

Indeed, one might argue, religion had not triumphed at all; rather, it had surrendered long ago, when Aquinas and others had relinquished to science the sacred right to define the Christian God and His heavenly realm. What science had given, therefore, it was now taking away.

In the ideas of Plato and Aristotle, science had given to Christendom a resplendent heaven, undefiled by terrestrial imperfections and run exquisitely by God. Now, though, in the theories of Copernicus, Brahe, Kepler, and Galileo, science was replacing it with a heaven despoiled by comets, ovals, and the beastly orbiting, spinning earth itself.

By corrupting the heavenly realm, science was now threatening to rob religion of that mysterious power and appeal it always had derived from being associated with lofty, godlike sublimity. In short, whereas religion was bringing science to its knees, science was bringing religion down to earth and dragging it through the dirt.

For its part, science now wished to be separated from religion. However, religion—having grown comfortable with its marriage and whose self-image was so greatly defined by its scientific spouse—wished desperately to remain wed.

After his trial, Galileo was placed under house arrest and left untormented for the remaining eight years of his life. Cataracts eventually blinded him, but to the end, he was able to see clearly that Plato's matchmaking had led to an unholy alliance.

In 1642, the beleaguered old Italian astronomer died, and by coincidence, Isaac Newton was born. In the years ahead, Newton would learn about the growing estrangement between science and religion and, in the end, bring about their permanent divorce.

VICI

The villagers were delighted and intrigued to hear that Isaac was on his way from Cambridge to be with his ailing mother. Over the years, they had kept well informed of the strained goings-on at the Newton-Smith manor; now the gossips wondered whether finally there would be a reconciliation.

To say that Woolsthorpe was proud of its most famous native son was a grotesque understatement; the tiny village venerated him and congratulated itself on having foreseen his notoriety; the fatherless child born on Christ's birthday was now a full-fledged *chaired* professor in the department of natural philosophy at the University of Cambridge.

The thirty-six-year-old had ascended quickly up the academic hierarchy because of several discoveries he had made. Any one of them alone would have been sufficient to secure Newton a place in history.

In a mathematical tour de force, for example, Newton had invented the calculus. Though in future it would become the bane of many a liberal arts student's college experience, seventeenth-century philosophers were thrilled to have been given a mathematical language that enabled them, for the first time in history, to describe the natural world with infinitesimal precision. (See "Between a Rock and a Hard Life.")

Also, Newton had expanded and refined Galileo's seminal work with metal balls by watching how objects moved in response to *any* force, not just gravity. Ultimately, he had been able to summarize their behavior with three simple truisms.

Truism #1. In a world where there are no forces to push things around, an object that is not moving will remain motionless forever, whereas an object that *is* moving will keep moving forever, along a straight line and at a constant speed.

Truism #2. In a world where there are forces to push things around, an object bullied by a force will always either accelerate or decelerate, depending on how the force is applied.

Truism #3. If two objects bump into each other, each will feel the force of the collision equally, but in opposite directions. (Years hence, many would paraphrase this by stating: "For every action there is an equal and opposite reaction.")

All together, Newton's achievements had made him renowned throughout the world, most especially at Woolsthorpe. He was an intellectual giant, and yet as the thirty-six-year-old pulled up to the stately manor, he trembled like a child at the thought of facing his bedridden mother.

When he entered the old house, he was greeted by his beloved uncle. The Reverend Ayscough was delighted to see Newton after all these years but also horrified to behold that the rumors he had been hearing from his old university acquaintances were true.

Newton looked terribly gaunt and distracted: During these past fifteen years, he had worked himself into a nervous breakdown from which he was still recovering. So far as the doctors had been able to ascertain, the breakdown had been triggered by a physical exhaustion caused by too much work and too little sleep, compounded by an emotional fatigue brought on by his constant feuding with colleagues.

The worst of it had started seven years earlier, in 1672, when Newton had been noticed by King Charles II and subsequently elected to the Royal Society of London. Being a member of this ultra-exclusive scientific academy was a singular honor for any natural philosopher, let alone one who had not yet turned thirty.

In keeping with tradition, the new inductee had submitted for the Society's consideration a report of his latest research. It was the scientific equivalent of a coming-out party, but it was to end in a disastrous falling out.

Up until then, most natural philosophers had believed that white light was absolutely pure, and that all the known colors were produced by its passing through some adulterating medium. For example, a little fouling produced red, a lot of it produced blue.

In their minds, that explained why white light passing through a glass prism produced all the colors of the rainbow. That portion going through the narrowest part of the prism's wedgelike shape produced red; that portion passing through the thickest part produced blue.

Newton had come to a completely different conclusion, however, after noticing that colored light passing through *any* part of a prism remained the very same color; red stayed red, blue stayed blue, and so on. Evidently, he surmised, it was colored light—not white light—that was pure and immutable. Indeed, white light appeared to be a composite of all the other colors, as evidenced by the fact that it produced the rainbow.

Excited by these extraordinary revelations, young Newton had thought them a grand way of introducing himself to England's elite Royal Society. Furthermore, buoyed by all this newfound collegial attention—which reminded him of that day, years ago, when he had been cheered for trouncing Arthur Storer—Newton had gone so far as to suggest immodestly that his discovery concerning white light was "the oddest if not the most considerable detection which has hitherto been made in operations of Nature."

The paper had been a hit, or so Newton had been led to believe. "I can assure you, Sir," the Society's diplomatic secretary Henry Oldenburg had effused, "that it mett with a singular attention and an uncommon applause."

Actually, however, put off by the self-important air of this young unknown and the audacity of his radical theory, a small number of Society members led by one Robert Hooke had greeted the publication with particular scorn and condescension. "As to his hypothesis," Hooke had snorted imperiously, "I cannot see yet any undeniable argument to convince me of the certainty thereof."

Of course, scientific criticism was quite routine and, in most cases, not meant to be taken personally. By questioning one another's theories with often brutal indifference to human feelings, natural philosophers intended to create a kind of intellectual jungle wherein only the fittest ideas would survive.

In this case, however, Hooke had been especially eager to discredit Newton, seven years his junior. Back in 1665, in a best-selling book titled *Micrographia,* Hooke had published an eloquent defense of the orthodox theory of colors, embellishing it here and there with ideas of his own. Hooke had become famous because of it—indeed, *Micrographia* had been his single greatest achievement—and he was not about to let it all be vitiated by the hairbrained hypothesis of some arrogant upstart. "For the same phaenomenon will be solved by my own hypothesis as well as his," Hooke had concluded defiantly, "without any manner of difficulty or straining."

Hooke's denunciations had rattled the reclusive and insecure Newton, calling up old memories of abandonment and rejection. He had tried to defend himself, to restate his results and reasoning as carefully as possible, but with no success: The critics could not be silenced.

Newton had become ill as a result, faulting Hooke most of all for poisoning his fledgling relationship with the Society. He had come to loathe this bully, and yet, rather than accelerating his resolve, the force of this newest kick to the stomach had caused Newton suddenly to withdraw from the only family he had ever embraced. "I will resolutely bid adieu to it eternally," he had lashed out bitterly, "for I see a man must either resolve to put out nothing new, or become a slave to defend it."

Though the bullies had cowed him, Newton had wished not to give them the satisfaction of knowing it. In his letter of resignation, therefore, he had pretended to be quitting the Society because London was too far from Cambridge for him to attend the meetings: "For though I honour that body, yet since I see I shall neither profit them, nor (by reason of this distance) can partake of the advantage of their Assemblies, I desire to withdraw."

After that, Newton had vowed never again to publish any of his work. During all these years, therefore, he had kept his ideas and experimental observations secret, scrawled in the pages of his notebooks; if his famous achievements had come to be known throughout the world, it was only because they had been leaked vaguely and incompletely via letters and word of mouth.

Never again had he sought to rejoin the Royal Society or society in general, for that matter. He even had given up any hope of ever joining up with Katherine Storer. In all this time, he had felt too insecure, been too studious, to give himself to the only young woman he had ever truly loved; in turn, she always had been too much of a lady to give herself to him. Now time had passed him by; another man had married her.

Walking into his mother's bedroom, Newton felt like the loneliest man alive: Already he had been rejected by his colleagues and by the fabled Cupid, and now it appeared he was about to lose this enigmatic woman who all her life had professed, if not shown, an undying love for him.

As he approached the large bed, Newton saw that his mother looked ashen and was barely able to speak, though she did manage a faint smile of recognition. He was moved; he had hated her most of his life, but now, faced with her extreme vulnerability, her mortality, something in his heart softened, and he wept like a baby.

She had not been much of a mother, but she was the one person whom he had secretly wished most to impress. He had been defiant with her, even cruel, but that behavior was behind him. Now, he pledged, his eyes awash in tears, his only desire was to show her how much he had loved her all along and had wished for her love in return.

Word of Newton's dramatic repentance spread throughout Woolsthorpe, and the villagers watched in wonder. According to one witness, Newton: "sate up nights with her, gave her all her Physick himself, dressed all her blisters with his own hands & made use of that manual dexterity for wch he was so remarkable to lessen the pain wch always attends the dressing."

Sustained by a lifetime's accumulation of unexpressed love, Newton hardly ate or slept. He was unfailingly at his mother's beck and call, one villager reported, "the torturing remedy usually applied . . . with as much readiness as he ever had employed it in the most delightful experiments."

Within a few weeks, his mother died and was buried in the village cemetery. In the aftermath of it all, Newton cursed himself for not having had a change of heart sooner than this, but the young natural philosopher also rejoiced at finally having discovered the feeling of a son's love for his mother.

In the days that followed, he remained in Woolsthorpe to help settle his mother's affairs and to reminisce. He walked through the pastures, rode to the windmill near Grantham—which was looking quite rundown now—and spent many hours with his uncle.

One warm evening, while he strolled through the garden, the moon began to rise, exactly as it had fourteen summers ago. Back then, Newton now remembered, he had done a calculation to show why the moon did not fall to the earth like some apple from a very, very high tree.

It didn't fall, he had figured, because the earth's gravitational force was being opposed by the moon's own centrifugal force; Newton chuckled when he recalled that, as a youngster, he had referred to it as the ring-a-ring o'roses force.

Now that he was older, he was more inclined to picture the situation in terms of a person being swung around on the end of a rope: The centrifugal force was what kept the rope taut, pulling with a strength that depended on just three things.

First, it depended on mass: A large grown-up being whirled around strained the rope far more so than a small child.

Second, it depended on the length of the rope: A very long rope produced a larger effect than did a short rope; for the person at the end of the rope, certainly, being swept around in a larger circle produced a far more dizzying experience.

Finally, it depended on speed: The faster a person was whirled around, the harder she strained against the rope and the more she had the feeling of being pulled away from the center.

Mathematically, if m stood for the person's mass, d stood for the rope's length, and T stood for how long it took to be whirled once around, then the centrifugal force the person felt was described by this simple equation:

$$\text{CENTRIFUGAL FORCE} = \frac{\text{constant} \times m \times d}{T^2}$$

In plain English: A large centrifugal force corresponded to a massive person or object being whirled around speedily on a long rope in a very short amount of time; that is, a large force resulted from multiplying a large m and large d and dividing by the square of a small T.

Conversely, a small centrifugal force corresponded to a light person or object being whirled around languorously on a short rope in a large amount of time; that is, a small force resulted from multiplying a small m and small d and dividing by the square of a large T.

As the garden swelled with the chirpings and croakings of its nocturnal inhabitants, a relaxed Newton harked back to how his mind had zeroed in on the T^2 in that formula. At first, he had been unable to remember where he had seen it before, but then it had come to him.

A century earlier, Kepler had argued that the planets whirled around the sun in orbits that obeyed a simple law:

$$T^2 = \text{constant} \times d^3$$

Admittedly, our moon was not a planet, Newton recalled having worried, but if it orbited the earth, as some said it did, then it, too, might obey Kepler's formula. If so, then he could replace his own formula's T^2 with Kepler's mathematical equivalent; namely, constant $\times$ d^3. Consequently:

MOON'S CENTRIFUGAL FORCE

$$= \frac{\text{constant} \times m \times d}{\text{constant} \times d^3}$$

$$= \text{new constant} \times m \div d^2$$

Back during that horrible plague-infested year of 1665, in other words, young Newton had come upon a most beautiful discovery. The centrifugal force the moon felt as it whirled around the earth depended on only two things (apart from the constant)—the moon's mass m and the length d of the imaginary rope connecting it to the earth.

That imaginary rope symbolized the pull of the earth's gravitational force. It yanked on the moon, and the moon's centrifugal force yanked back in the opposite direction. The result was a cosmic standoff, young Newton had reasoned, which explained why the moon, instead of falling down or pulling away, circled round and round in a kind of eternal holding pattern.

Filled with nostalgia, Newton now recalled the climactic moment of that fateful night, when he was but twenty-three years old. If he were correct about the standoff, he had concluded, if the strengths of the two opposing forces were equal, that meant they obeyed the same mathematical equation:

EARTH'S GRAVITATIONAL FORCE
= MOON'S CENTRIFUGAL FORCE

$$= \text{constant} \times m \div d^2$$

That is, the earth's gravitational pull weakened the farther away it was from the earth—it weakened with the *square* of the distance (i.e., a smaller and smaller force resulted from dividing m by a larger and larger d^2).

For instance, an apple *two* times as far away from the earth would feel one-fourth the pull. (In other words, the force was diluted by four, the square of two.) An apple *three* times as far away would feel one-ninth the pull, and so forth. By the time one got as far away as the moon, the earth's pull would be feeble indeed, but it would still exist.

As far away as one could imagine, in fact, the earth's pull would still exist. Its strength never vanished completely; it merely faded away as one traveled farther and farther from the earth, toward infinity.

That last assertion, Newton now realized better than he had back then, was a frightfully heretical concept. Here was a perfectly reasonable argument for thinking that the earthly realm might extend to the farthest reaches of the universe, in direct contradiction to Aristotle's belief that it stopped just short of the moon.

As Newton picked himself up to return to the house, he looked up at the sky one last time and wondered what the heavens were trying to tell him. He was not an avid astrologer by any means, but like Kepler, he had always been inclined to believe in the interconnectivity of the universe's two realms.

God, he believed, intervened in our daily affairs out of necessity. Indeed, Newton mused, as he climbed the stairs to his bedroom, one could think of life as being another kind of cosmic standoff: Ever since Adam and Eve had bitten into the apple, God's redemptive presence had been the only thing keeping this imperfect world from falling into ruin.

Coincidentally, as Newton fell asleep that night thinking afresh about the tug of war between the forces of heaven and earth, people in London were being kept awake by a similar struggle involving Roman Catholics and the English government.

Having so recently been ruled by overly strict Puritans, the English regarded any and all non-Anglican zealots with unmerited suspicion and malice; in a word, they were quite jumpy. Most recently, for example, the Pope had been rumored to have recruited the king's brother James II in an effort to assassinate Charles II; in the ensuing paranoid frenzy, many innocent Roman Catholics had been massacred.

When he returned to Cambridge, furthermore, Newton returned to a university that by law excluded from its faculty anyone who did not sign a loyalty oath. According to this so-called Test Act, in fact, no one could hold *any* public or military position who refused to receive communion according to the tenets of the secularized Church of England.

English natural philosophers were among the most enthusiastic supporters of the Test Act, which they saw as a timely sanction against the Roman Catholic church's continued persecution of science. After all, they noted, the Vatican still kept Galileo's writings on its reprehensible list of prohibited books. (And would continue to do so until October 31, 1992!)

In seventeenth-century England, religion was less wedded to science and, therefore, more tolerant of science's mercurial opinion of God's creation. Conversely, science was more tolerant of religion. Indeed, many of Newton's contemporaries were devout servants of both realms.

As theologians, they read the Bible and critiqued one another's interpretations of it. As natural philosophers, they did experiments and critiqued one another's theories of how best to explain the results. Among Anglicans, one might say, science and religion had separated. They now lived in their own houses, and to the extent they interacted, they tried to get along, even to reconcile their widening differences.

Many of Newton's colleagues, for example, were trying to reconcile the laws of science with the biblical account of the Flood. It would take them years to finish, but after lengthy and controversial calculations, they would eventually conclude that the Deluge had begun precisely on November 28, 2349 B.C., when a low-flying comet created huge fissures in the ground, allowing water to escape the oceans and flood the earth.

Newton himself wore two hats: If he wasn't picking apart the meaning of the obdurate predictions contained in the Book of Revelation, he was trying to change iron into gold. Though he wasn't much of an astrologer, he was becoming quite proficient in alchemy, the forerunner of modern chemistry.

The direction of Newton's thoughts was changed completely, however, by a letter he received from his old nemesis, Robert Hooke. Unbeknownst to Newton, Hooke had come to admire his achievements from afar, albeit grudgingly and jealously, and now wished to have Newton's opinion about a new idea.

Hooke had given Kepler's oval-shaped orbits a great deal of thought over the years, the letter explained. As a result, he had concluded the orbits probably were caused by a gravitational force that weakened with the *square of the distance* from the earth!

Hooke had come to that idea, he explained, by imagining the earth to be like a source of light—a candle, one might say. A century ago, Kepler had discovered that brightness diminished with the *square of the distance* from the light source: A candle two times as far away appeared one-fourth as bright; a candle three times as far away appeared one-ninth as bright; and so forth.

Perhaps, Hooke conjectured in his letter, earth's gravity also waned with distance, just like a light's brightness. If so, Hooke concluded, "the Attraction always is in a duplicate proportion to the Distance from the Center Recriprocall"—in other words, the gravitational attraction always diminishes in proportion to the square of the distance from the center of the earth.

As Newton read the letter, he smirked: The bully had lucked

upon the truth. But no matter. If only the odious little man knew how far behind he was in his thinking. Fourteen years ago, Newton had actually calculated the result at which Hooke was now only guessing.

In the days ahead—though he had dismissed Hooke's letter as so much child's play—Newton began to wonder about the loose ends that had been left untied by his own efforts back in 1665, chief among them being this question: What was the *cause* of earth's gravitational field? The philosopher's treasured Principle of Sufficient Reason demanded an answer.

He dismissed Descartes's tornado theory, because if it were true, the apple in the garden would have *spiraled* down to the earth; instead, Newton had noticed carefully that things fell *straight* down. It was as if the *center* of a falling object were being yanked to the *center* of the earth, not off to one side or another.

At that point, Newton began to wonder: What would happen if the earth was whittled down to the size of a tiny particle at its center and, likewise, the apple was whittled down to a tiny particle at its core? Would the tiny apple-particle fall toward the tiny earth-particle? He could think of no reason why not, whereupon he struck on the idea that would lead to his famous equation.

Everyone was accustomed to thinking of the apple falling toward the earth, because the apple was so much smaller than it. By reducing the situation down to two equally sized particles, however, it became implausible to keep believing that the apple-particle would fall while the earth-particle would just sit there unmoved.

It was more reasonable, more equitable, to suppose that the two particles fell toward one another. In other words, what we possessively referred to as *earth's* gravity did not belong to the earth exclusively; gravity was a force of attraction felt mutually by *all* particles of matter.

These newer revelations did not disqualify the gravitation equation Newton had first found as a young man, but they did require it to be amended slightly. The original equation had been formulated with the idea that earth's gravity was a unilateral force, so the equation contained a reference to only the mass of the object being attracted to the earth; in recognition of gravity being a mutual force, the equation needed an explicit reference to the mass of the earth being attracted to the object.

Alongside the m, which referred to the object's mass, therefore, Newton inserted an M, which represented the earth's mass. That way, both object and earth held an identical place in the revised equation, in keeping with gravity's perfect reciprocity:

EARTH'S GRAVITATIONAL FORCE
$$= \text{constant} \times M \times m \div d^2$$

In plain English: Between the earth and massive objects close to it, the force of attraction was very strong and irresistible; between the earth and tiny objects far away, the force was quite weak. In short, the earth and any other object were attracted to one another with a force whose strength depended on: the distance between their centers, their two masses, and some constant number.

In years to come, scientific experiments would determine the value of that number with enormous accuracy. In remembrance of the man who first made reference to it, furthermore, it would come to be called "Newton's gravitational constant" and designated by the letter G. Ultimately, therefore, it would take a little less space to write out the equation:

EARTH'S GRAVITATIONAL FORCE $= G \times M \times m \div d^2$

In the most general terms imaginable, Newton's equation expressed the gravitational force between any two objects; the letters m and M could stand for the mass of the moon and Jupiter, or a comet and the sun, or any other pair of bodies.

Gravity was a force of attraction that was felt mutually by *all* particles *everywhere* in the universe; in short, Newton concluded, gravity was the glue that held everything together.

After all these centuries, Aristotle's large-minded theory of the heavens had been pulverized by Newton's tiny-minded theory of gravity. According to this new vision, the universe was not segregated into two separate realms; there was only one universe, ruled not by some divine monarch but by one very earthly gravitational equation.

Much of what the universe had been, was, and would be, Newton had disclosed, was the outcome of an infinity of material particles all pulling on one another simultaneously. If the result of all that gravitational tussling had appeared to the Greeks to be a cosmos, it was simply because the underlying *equation* describing their behavior had itself turned out to be every bit a cosmos—orderly, beautiful, and decent.

In 1682, as if in celebration of Newton's remarkable discovery, the heavens produced a comet over the skies of London. It was not a very bright comet, however, perhaps because Newton was not in a partying mood.

After all these years, this brilliant and successful philosopher had not gotten over his hurtful experience with the Royal Society. Excited though he was by his discovery, he feared being criticized for it. Therefore, he decided, he would not publish the equation.

Some years later, Newton received yet another letter from Hooke, who was now secretary of the Royal Society. Hooke had heard about Newton's gravitational equation and wanted to make certain that Newton agreed it was he, Hooke, who had first come up with the "distance-squared" theory; as proof, he reminded Newton of the letter he had sent years earlier describing the idea.

Newton was livid. "I am not beholden to him for any light into this business," he protested vehemently in a letter to a colleague, "but only for the diversion he gave me from my other studies to think on these things."

The petty tyrant was attempting to cow him again, Newton steamed, but it wouldn't work. This time, he would respond the way he had as a grammar-school student in Grantham: He would beat this tormentor senseless, once and for all.

In the years that followed, Newton set aside his alchemical and religious studies and devoted himself to exhuming every discovery he had ever made. He searched through all his papers, even his childhood notebooks, refining conclusions and redoing calculations.

Newton did all this work himself, but he was encouraged every step of the way by an astronomer named Edmund Halley. After years of vain efforts, Halley had been overjoyed to hear of Newton's equation of gravity; with it, he finally had been able to make sense of cometary behavior.

In fact, after hundreds of hours of searching back into the historical records, Halley had concluded that the recent 1682 comet had been the same one Kepler had seen in 1607 and that others had observed many times before then. Using Newton's equation, he had figured out that the comet was in *orbit* around our planetary system, passing by the earth roughly every seventy-six years; it would reappear, he predicted, in 1758.

This was a far-fetched forecast, because heliocentric astronomers since Kepler had come to believe that comets traveled along straight lines: They passed by earth once, they believed, never to reappear again. "Should the comet return according to our prediction," Halley stated imperially, "posterity will not refuse to acknowledge that this was first discovered by an Englishman."

With financial help from Halley and the blessing of the Royal Society itself, Newton finally came clean with the world from which he had lived apart for nearly all his life. In 1687, he published his life's work in three volumes and titled it *Philosophiae*

Naturalis Principia Mathematica (Mathematical Principles of Natural Philosophy).

The monumental publication stunned his English colleagues and, with its powerful marriage of mathematics and experimentation, transformed natural philosophy into natural *science;* yet it was missing something. The crafty orphan from Woolsthorpe had decided to withhold from this magnum opus any mention of his ideas concerning light; he would not publish those until the bully Hooke had died—which would not happen until 1704—thus guaranteeing himself the pleasure of having had the last word.

In one way, by discrediting the idea of a segregated, two-realm universe, Newton's scientific revolution crushed the rebellion Plato had begun two thousand years earlier. In another way, however, it represented the utter fulfillment of Plato's wish that humanity "cast off the superstitious fear of prying into the Divine."

What Plato hadn't foreseen was that in the process of helping us cast off our fears, science would help us cast off our gods. Earth's gravity, Newton had demonstrated, extended to the moon and beyond; indeed, there was no place in the universe that did not feel its influence, however remote it might be.

Consequently, there was no place left uncorrupted in the universe for God to dwell. He had been crowded out of our picture of the universe by gravity's infinite reach. For the first time in Western history, the heavens had been completely despoiled; God's perfect existence had been purged ignominiously from our scientific theories.

The historic betrothal Plato had arranged had now ended in complete ruin: As a result of our investigating the heavens, science had become irreligious and religion had become unscientific. It was a momentous parting of the ways, and though Newton was the main person responsible for the troubled marriage's final breakup, he had a surprise accomplice of sorts.

In 1688, only months after Newton's revolutionary publication, the English had decided they had had enough of their new king. James II had succeeded Charles II only three years ago, but already his flagrant Catholicism had brought his country to the brink of another civil war.

To keep that from happening, English politicians of all faiths had cooked up a scheme, which began with their sneaking into the country a Dutch prince named William of Orange and his consort, the king's Protestant daughter, Mary II. The next step now was for Parliament to declare that James II was no longer the king of England.

Predictably, the king responded by reminding England that he ruled by divine right, just as his predecessors had done. He had been appointed by God Himself to lead the English people, and it was sacrilege for any secular institution to presume to countermand His authority.

At the sight of William leading a large army into London, however, James quickly gave way and fled the country. It was called the Glorious Revolution, because from then on, for the first time in history, Parliament would have the undisputed authority to appoint England's kings and queens.

With that, the Western world had begun to expurgate God from its government as well as its cosmology. Politically and scientifically, the influence of the earthly realm had vanquished the age-old authority of the heavenly realm; God and his representatives were no longer wanted or needed to govern the English people or Newton's cosmos.

State separated from church; science divorced itself from religion. These were historic and enduring disconnections. Even three centuries hence, modern Western civilization would show the effects of being the offspring of divorced parents: Its people would live in a scientific and political world without God and a religious world without science—the remarkable legacy, one might say, of an apple from Woolsthorpe and a prince from Orange.

EPILOGUE

The 1960s were a time when it seemed nothing could go right for the United States. It was the era of the Vietnam War, of leaders being assassinated, of violence raging in the streets; it was a time of great pessimism.

It was not surprising, therefore, that in 1969, many people thought the idea of going to the moon was impossible. Some were skeptical for technical reasons: How could we transport ourselves to something that was one-quarter of a million miles away, let alone land on it and then return safely?

Others were doubtful for religious reasons. The earth's gravity might extend into the heavenly realm, they conceded, but earthlings themselves would never do so—would never plant their dirty feet on the moon or any other heavenly body.

The doubters notwithstanding, the United States had pressed ahead, under the leadership of the National Aeronautics and Space Administration. The forerunner of NASA had been formed back in 1957, immediately after the Soviets had launched the world's first satellite, and now it was well on its way to planning the world's first round trip to the lunar surface.

Politically speaking, NASA was acting in response to President Kennedy's 1961 State of the Union challenge: "I believe that this nation should commit itself to achieving the goal before this decade is out of landing a man on the moon." If successful, the United States would score a stinging Cold War victory against communism.

Genetically and scientifically, however, NASA was responding to that irresistible human urge to explore the unknown. The space agency was racing to beat the Soviets, yes, but it was also trying to fulfill a visceral desire first articulated by the astronomer Johannes Kepler in *Somnium* (Dream), history's first work of science fiction.

Published posthumously in 1634, *Somnium* had described a boy journeying to the moon with the supernatural aid of a friendly demon, conjured up by the boy's witch of a mother. The story was quite unbelievable, but it had survived to infect other writers with the dream of going to the moon, most notably a Frenchman named Jules Verne.

In his 1865 novel, *From the Earth to the Moon,* Verne had described a trip to the moon in prophetic detail. According to the popular author, three men had made the long journey inside a huge aluminum bullet, fired from a 900-foot-long cast-iron cannon located in Tampa, Florida.

Now, a century later, NASA was planning to send three men to the moon traveling inside what amounted to a giant titanium bullet, fired from a launch pad in Cape Canaveral, Florida, one hundred miles directly east of Tampa. The astronauts would not be shot from a cannon, but they would ride atop a 363-foot-long liquid-fueled rocket, the *Saturn V.*

In preparation for that trip, NASA had sent a group of astronauts, including Neil Armstrong, to the Lowell Observatory in Flagstaff, Arizona, to have their first close-up look at the moon. They could have gone to any number of observatories in the United States, but it was especially significant that NASA had chosen this one.

The observatory had been founded in 1894 by Percival Lowell, a well-to-do eccentric who had wanted a telescope to search for life on Mars. Though he never did end up finding any "little green men," his observatory had become one of the country's most prestigious facilities for studying the solar system.

When the Lowell Observatory had first opened, people believed the solar system consisted of seven planets (in addition to the earth). There were the five planets known to Copernicus, plus two more—Uranus and Neptune—that astronomers had discovered in the years since.

Astronomers had noticed, furthermore, that Uranus's orbit was not perfectly oval, in violation of one of Kepler's laws. That had led many of them, including Lowell, to blame the aberrations on the gravitational pull of a nearby planet not yet discovered.

Armed with nothing but Newton's gravitation equation and his own brand-new telescope, Lowell had predicted the probable location of this hypothetical planet. He had not lived to see it happen, but in 1930, his assistant Clyde Tombaugh had found the planet only six degrees away from where Lowell had said it would be; subsequently, astronomers had named it Pluto.

Now, in 1969, Newton's equation was poised to play a similarly crucial role in sending astronauts to the moon. Indeed, if NASA succeeded in its mission, it would be only because Newton had given us the mathematical wherewithal to find our way there.

Using Newton's equation, astronomers over the years had calculated the moon's orbit so precisely that NASA engineers were now able to know exactly where their lunar target would be at any moment in time. By calculating the rate at which earth's gravity diminished at any point along the way to the moon, furthermore, NASA had been able to determine what size rocket was needed for the job—one twice as tall as the Statue of Liberty, as it turned out!

It was in order to give their rockets an extra 5 percent boost, furthermore, that NASA had originally chosen to launch them from Cape Canaveral. There, close to the equator, the effect of the earth's spinning was felt more so than anywhere else in the country; that is, objects were whipped around with the greatest centrifugal force at the equator, because the equator was farthest from the earth's axis. When a rocket took off from Cape Canaveral, therefore, it was like being slung off the edge of a fast-spinning carousel.

To take full advantage of this earthly boost, NASA usually preferred to launch rockets eastward, *with* the earth's spin, not against it. Fortunately, they were able to do this safely, because immediately to the east of Cape Canaveral was the Atlantic Ocean and only a few sparsely populated islands.

Upon first hearing of President Kennedy's challenge, engineers had realized that it was not going to be as simple as launching a rocket and pointing it toward the moon. To calculate the best route to the moon, therefore, NASA had created the Mission Planning and Analysis Division (MPAD), based at its Mission Control Center in Houston, Texas; in 1969, at its height, MPAD consisted of nearly a thousand scientists and engineers.

Theirs was a dreadfully complicated task, because it required applying Newton's equation to three objects simultaneously—earth, moon, and spaceship—not just two. It was what scientists referred to as a *three-body problem:* As the spaceship sped along on its journey, its distances from the earth and moon would be changing constantly; consequently, the gravitational pulls between it and the two bodies would be changing constantly.

Keeping track of all that, predicting the net effect of three objects pulling on one another, was impossible to compute exactly. In applying Newton's equation to three-body problems, the best one could hope to do was *approximate* an answer, and even that much could not be done without the aid of computers.

Equal to the challenge, NASA had provided MPAD engineers with state-of-the-art IBM computers. They occupied the entire first floor of the Mission Control Center, and for the past several years, they had operated twenty-four hours a day, seven days a week, and fifty-two weeks a year; on the eve of NASA's first attempt to land on the moon, the engineers and computers had calculated the safest and cheapest way to get there.

The astronauts were to travel to the moon and back along a route shaped like a figure-eight; any other shape, it turned out, was either more dangerous or more wasteful of rocket fuel. Furthermore, by following such a smooth and simple trajectory, the astronauts could whip around the moon and return safely to earth in case they had to abort the mission at the very last moment; in such an emergency, Newton's equation had predicted, no fuel would be required, because the moon's gravitational pull would automatically sling the spaceship around and onto the return leg of the figure-eight.

On the morning of July 16, NASA engineers had done everything they felt was necessary to prove the skeptics wrong. They had confidence in all their calculations; nevertheless, when the critical moment arrived, they held their breath as the three astronauts lifted off amid a fiery explosion and billowing cloud of steam.

The giant rocket inched slowly upward, struggling against the unrelenting force that had held us captive on this earth for all of our species' existence. As the rocket thundered its way into the clouds, it began to spin like a bullet; long ago, scientists had figured out that putting a spin on a fast-moving projectile kept it from wobbling off course—the same reason a child's spinning top stays upright.

At first, astronauts Neil Armstrong, Buzz Aldrin, and Michael Collins sped toward the moon at 25,000 miles per hour, the running start needed to pull free of the earth completely. For days, fighting earth's gravity was like traveling uphill. Two-thirds of the way there, however, 190,000 miles away from the earth, the spaceship started to speed up, as if it were going downhill: The astronauts had reached the point at which the moon's gravity was stronger than the earth's.

On July 20, at 3:18 P.M., Houston time, as more than 600 million people watched the lunar lander come to rest on the moon's rock-strewn Sea of Tranquillity, NASA engineers breathed a loud sigh of relief; the *Somnium* had come true. A short while later, as the world watched Neil Armstrong take his first step onto the moon, those same NASA engineers cheered: "That's one small step for man," Armstrong intoned, "one giant leap for Mankind."

Had he been alive, no doubt, Newton would have cheered right alongside the men and women who had taken such spectacular advantage of his equation. It was a historic moment, made possible by a historic equation.

For the posthumous child from Woolsthorpe, moreover, it was a posthumous honor befitting a man who in his last years of life had finally discovered the familial love for which he had longed so desperately. Following his discovery of the universal gravity equation, Newton had been elected president of the Royal Society, appointed a member of Parliament, and knighted by Queen Anne, daughter of England's last heavenly monarch, James II.

During that time, Sir Isaac had hobnobbed with aristocracy and entertained guests from high society in his posh London apartment. He had never married, but a niece named Catherine Barton acted as his hostess. Her beauty and intelligence beguiled even the great French philosopher and playwright François Marie Voltaire.

The world had become Newton's family, with a few notable exceptions. Having laid Robert Hooke to rest, literally, Newton had become involved in an all-out feud with a German philosopher named Gottfried Wilhelm Leibniz, who had claimed credit for discovering the calculus. (See "Between a Rock and a Hard Life.")

During that time, furthermore, he had reflected on his most famous equation and its stupendous consequences, conceding that "we have explained the phenomena of the heavens . . . by the power of gravity, but have not assigned the cause of this power."

Ultimately, he had insisted, God was the cause of it all. "This most beautiful system of the sun, planets, and comets," Newton believed earnestly, faithfully, "could only proceed from the counsel and dominion of an intelligent and powerful Being."

Aristotle had been wrong to think of God as being confined to a heavenly realm separated from the earth, Newton had concluded, and now it was just as wrong for his younger contemporaries to surmise that because of the thoroughgoing gravitational "corruption" of that perfect domain, God had been ostracized from the universe.

Instead, the Creator always had been, was, and would be everywhere throughout His Creation, even in the tiny-most particle of apple and earth. "He is eternal and infinite; omnipotent and omniscient," the aging natural scientist had held passionately; "his duration reaches from eternity to eternity; his presence from infinity to infinity."

Newton died in the wee hours of the morning on March 20, 1727, and was buried in Westminster Abbey, the church where nearly all English monarchs since William the Conqueror had been crowned and only the most famous of the famous were interred. His casket had been borne by nobility: three dukes, two earls, and the Lord High Chancellor.

He had been the first scientist to be honored so lavishly, and yet, even if he had been alive to boast about it, he most certainly would not have done so. Newton had died a man satisfied that all the bullies of the world had been put in their place by the great esteem and affection with which the world held him. That had permitted him to become humble: "If I have seen further," he had said at one point, "it is by standing on the shoulders of Giants."

Luckily for us, Newton had taken us along for the ride. With his marvelous equation, he had hauled us up on *his* shoulders, and in 1969, as Neil Armstrong walked about in the celestial realm, we were astonished by what we saw and felt.

The experience was grand and godlike, yet in the end, disquieting. We had conquered the heavens, but in that year when we witnessed firsthand the empty vastness of the purely scientific cosmos, we felt meeker and lonelier than at any other time in human history.

$$P + \rho \times \tfrac{1}{2}\, v^2 = \text{CONSTANT}$$

Between a Rock and a Hard Life

Daniel Bernoulli and the Law of Hydrodynamic Pressure

Destiny has more resources than the most imaginative composer of fiction.
—FRANK FRANKFORT MOORE

A s the homing pigeon flew over the houses, thirty-four-year-old Daniel Bernoulli stopped to watch. How wonderful it must be to fly, he thought, and how swiftly a bird was able to go from here to there; his own recent trip home from Russia had taken an entire two months, traveling by horse-drawn coach.

As he turned away and began to gather the mail, Bernoulli's heartbeat quickened when he spotted the letter from Paris; no doubt, he guessed, it contained the contest results. Oddly, though, it was addressed to him *and* his father Johann; they both had entered the competition, but they had submitted separate essays.

Every year, the French Academy of Sciences challenged the public to solve a technical problem of some importance. This was not the only contest of its kind—scientific institutions in several European countries did the same thing—but this was one of the oldest and most prestigious in the world. For the past sixty-eight years, since its establishment by King Louis XIV in 1666, scores of engineers, mathematicians, and laypeople had vied for the money and prestige that came with winning.

So far, young Bernoulli had entered the contest a total of four times and already had won once. He was mathematically gifted in all subjects, but he especially loved tackling problems involving fluids. Scientifically speaking, fluids included not only all kinds of liquids but also gases and any other pliable material that was not completely solid.

Fluids fascinated the mathematician in Bernoulli, because they were complicated enough to be challenging yet simple enough to be scrutable. Furthermore, fluids were so much a part of day-to-day living that studying their actions was a useful and relevant thing to do—and the time seemed right.

In the seventeenth century, Isaac Newton had successfully described the behavior of *solid* objects. And in the nineteenth century, scientists would discover the laws of genetics, evolution, and psychology that governed the activity of *human beings*. In between those two centuries lay Bernoulli's century, a time destined to belong to fluids, whose complexity lay somewhere between a solid rock and human existence.

Bernoulli had always dreamed of becoming the Newton of his time, of being the first to discover the laws that steered the movement of fluids. That was why, over the years, he had made it a point to enter the French Academy's contest whenever it involved a fluids problem: It was an invaluable opportunity to exercise, and to showcase, his precocious talents.

Now, as he ripped open the envelope, he drew a deep breath: He had just returned to Basel after having spent the past eight years at the Russian Academy of Sciences. What a nice homecoming present it would be if his essay had been declared the winner this year.

After pulling the letter out of the envelope, Bernoulli unfolded it and began to read. It was, as he had suspected, the announcement of this year's contest results, but what he saw made him gasp.

For the rest of the afternoon, the young man waited eagerly for his father to arrive. He had decided not to track him down at the university, knowing that the famous Professor Johann Bernoulli generally became angry at anyone who dared to disturb him while he worked.

When his father arrived that evening, the young Bernoulli greeted him with the letter, without saying a word about what it contained. Quizzically, the stern-faced professor took the missive and read for himself that the Academy had decided to award this year's first prize to both father *and* son.

Young Bernoulli, who could not contain his excitement much longer, anticipated that he and his father soon would embrace with elation; but it was not to be. In a matter of seconds, young Bernoulli sensed that something was amiss.

His father was reacting not with a shout of jubilation but with a joyless silence. Worst of all, when he was done reading it, he crushed the letter in his fist and glowered at Bernoulli, erupting with an effusion of terrible accusations.

At first, Bernoulli was paralyzed with confusion. But slowly he began to comprehend the reason for this ghastly turn of events.

The elder Bernoulli, who years ago had introduced his son to mathematics and taught him many of the basic ideas and techniques that underlay their pair of prize-winning essays, was furious that the young man should now be considered his equal. He denounced the Academy for not distinguishing the master from the pupil and derided his son for not giving him proper credit.

As his father's rage intensified, Bernoulli himself became angry. Being away from home these past eight years, he not only had practiced and perfected the ideas and techniques his father originally had taught him, he had enhanced them in his own way, without anyone's help.

It was as if he had learned from his father how to run the farm machinery but then, by himself, had gone on to plow and plant his own field; now, properly enough, he was harvesting the rewards of his own labor and skill. What's more, the young man shouted impudently, *his* essay was better than his father's!

As evening fell and the city grew quieter, the hateful sounds coming from the Bernoulli household grew louder. The two men roared at each other, availing themselves of the opportunity to vent old, pent-up grievances. By the time their bitter clash reached its climax, the original quarrel about the Academy prize had been superseded long since by passionate complaints about filial disrespect and paternal jealousy.

In the end, the elder Bernoulli demanded that his ungrateful offspring leave the house, screaming that he could not tolerate living with such a miscreant. Bernoulli, in the midst of the escalating tension, had feared it would come to this. Now, as he heard himself being evicted, he regretted immediately many of the things he had said to his father.

Young Bernoulli had always been proud of having descended from a family of distinguished mathematicians. He was the son of a man who was arguably *the* most renowned mathematician alive and the nephew of a similarly famous mathematician. In fact, Bernoulli men had dominated mathematics for the last fifty years, a pedigree the likes of which had never been seen before and might never again.

Bernoulli was sad this great old family tree of his suddenly was not faring too well; he feared being severed from his roots, perhaps forever. Still, he was too angry to apologize or to sleep under the same roof with a man whom he had long admired but now mistrusted.

It took him less than an hour to pack his belongings, and as he walked out the door, he paused to look back. He had been born here, and he would miss living here . . . and if truth be told, he would miss the spirited conversations he recently had been having with his father about the latest theories concerning fluids.

Now, more than ever, dealing with fluids seemed so much more appealing to Bernoulli than dealing with people. With fluids, at least, there was some hope that their behavior would turn out to be predictable. By contrast, people's behavior seemed hopelessly unfathomable; for example, Bernoulli thought with a shrug, who could have predicted what had happened today?

As the young man stepped out into the cool autumn darkness, he wondered where he would spend the night. For Bernoulli, regrettably, this was only the beginning of what was to be a steady and tragic decline in his personal fortunes; but it would not end in total ruination.

In the course of his life, the young mathematician was to come upon a magical equation that would reveal the secret of flight. As a result of that, his scientific reputation would soar . . . and so would the mind, body, and spirit of the human species.

VENI

Unlike the medievalists immediately before them, Renaissance philosophers were disinclined to invoke supernatural explanations for the puzzling phenomena they saw and heard around them. Instead, they gradually readopted the ancient Greek attitude that, for every mystery that existed in the natural world, there was a *down-to-earth* explanation.

Indeed, Renaissance scholars went so far as to say that, by knowing the rational laws of Nature, they could actually *foretell* the future. For example, sixteenth-century astronomers argued, if only they knew the scientific laws of heavenly bodies, they easily could predict a planet's orbit.

Already, *astrologers* were claiming to have the ability of prophesying a *person's* fate; but their mystical methodology—though based on sound astronomical observations—was regarded with suspicion by science. Someday, pundits asserted, by discovering the bona fide scientific laws of human nature, they truly would be able to predict human behavior.

As the seventeenth century unfolded, however, science encountered reasons both to hope and to despair that it would ever realize its bold objective. On the one hand, in 1687, when Isaac Newton published his astonishing discoveries governing the behavior of inanimate objects, it seemed certain that similar discoveries soon would be made regarding the behavior of *animate* objects.

On the other hand, the colorful saga of one particularly prominent European family, the Bernoullis, seemed to bode ill for science's great ambition. Indeed, the Bernoullis seemed to prove that while people can be predictable in certain ways, it was sheer folly to hope their overall fates ever could be divined scientifically.

The Bernoullis' eccentric and quarrelsome tale began in 1622, when Jacob the Elder fled to Basel, Switzerland. He had been born in Belgium and was a diehard Huguenot persecuted mercilessly by the Catholic majority.

Basel's reputation for religious tolerance was renowned the world over; indeed, the Huguenots' spiritual leader John Calvin himself had emigrated there from his native France, following the tumultuous sixteenth-century Reformation. Like Martin Luther before him, Calvin had believed that in God's Master Plan, each of us played a predetermined role.

In Jacob Bernoulli's eyes, therefore, it had been his *destiny* to: prosper in Basel, marry three times, and father only one child. Toward the end of his life, the old patriarch regretted not having had more offspring, but he need not have worried; his lone son Nikolaus was destined to sire an extraordinary dynasty, as tempestuous as it was talented.

After marrying, Nikolaus and his wife gave birth to a total of a dozen children, of whom only four would survive into adulthood.

Two of the survivors would become world-class mathematicians—Jakob, born in 1654, and Johann, born in 1667—though at first their father did not see it quite that way.

When the sons were young, father Nikolaus was certain Jakob's brooding intellect meant he was to become a theologian and that Johann's naturally thrifty ways indicated he was to become a merchant. Acting on those beliefs, consequently, the devout Huguenot demanded that each son prepare himself for his predetermined role in life.

Feigning obedience, Jakob went on to earn a master's degree in philosophy at the University of Basel and a license in theology. But covertly, he pursued his real love, physics and mathematics. "Against my father's will," he confessed in his diary, "I study the stars."

Thirteen years younger than Jakob, Johann behaved dutifully. He acquiesced to become an apprentice in the family's prosperous spice and drug business, but ended up performing so miserably that the disgusted father was forced to recant his original premonition.

God's plan was now more clear to him, Nikolaus announced one day: Johann was meant to be a doctor. It was a profession still related to the family business, and it would provide a handsome living.

Obediently, sixteen-year-old Johann entered the University of Basel and earned his medical license. During that time, however, he conspired with his older brother Jakob to learn the language of numbers. "I've now turned to mathematics," the teenage Johann admitted in his diary, "for which I feel a special joy."

Coincidentally, Johann's clandestine conversion took place around the time German mathematician Gottfried Wilhelm Leibniz came out with a landmark paper announcing his discovery of the calculus. It was a revolutionary new kind of mathematics that had been developed independently—though not yet published—by the Englishman Isaac Newton.

Printed in 1684, Leibniz's article did not elicit any immediate response, simply because very few people in the world could comprehend it. The author, with characteristic arrogance, had not tried very hard to explain his discovery, presumably because he wanted to remind people of how much smarter he was than they.

The Bernoulli brothers, too, were unable to make much sense of Leibniz's paper, despite their dogged efforts to do so. They even wrote to the great mathematician, begging for help, but did not receive so much as a reply.

Undiscouraged, they persevered, until one day, as if by some miracle, Jakob suddenly understood everything. Thereafter, he shared his epiphany with Johann, so that together they could explore the subtle minutiae of Leibniz's monumental achievement.

It all hinged on something called the "infinitesimal," Jakob explained, an imaginary whit so infinitely tiny as to be devoid of any complexity; it was tinier even than the tiniest imaginable speck of paint on one of Vermeer's variegated masterpieces.

Here then was the crystal ball philosophers had been wanting all these years; by reducing complex processes down to their infinitely tiny, infinitely simple parts, Leibniz's calculus gave science a way of predicting the unpredictable—including, perhaps, human behavior!

According to Leibniz, with the calculus, the seemingly unpredictable process of selecting a lottery winner—whether it involved the tossing of dice or drawing of lots—could be broken down mathematically into a sequence of infinitely simple events, each of which was easily predictable. In the end, merely by adding up the outcomes of all those infinitesimal events, one could divine the result of the entire process.

Leibniz's novel mathematics appealed instantly to the Bernoulli brothers' Calvinist training, insofar as it seemed to validate their belief in predestination. If God knew in advance what people's futures were to be, then the calculus was the technique they could use to read God's mind.

For three years, Jakob and Johann struggled excitedly and secretly to increase their fluency in this wondrous new mathematics; then, much to their surprise, they received a belated reply from Leibniz. Writing back to him immediately, the young erstwhile theologian and merchant exclaimed their progress. From that moment, they enjoyed the very rare privilege of corresponding regularly with the lofty codiscoverer of the calculus.

Exceedingly less enjoyable for them was the day their father discovered their deceitful behavior; immediately, Nikolaus Bernoulli demanded they find well-paying jobs. He no longer cared what kind of jobs they found, he shouted, but he absolutely had no intention of subsidizing their pursuit of so unprofitable a preoccupation as *mathematics*.

Shortly thereafter, despite his father's invectives, Jakob was hired as a professor of mathematics at the University of Basel; there, in the years to come, he became famous for his success in using the calculus to solve complicated problems in every known field of science, from chemistry and cosmology to engineering and economics. In the process, though, he revealed himself to be a slow, methodical thinker—the proverbial tortoise in Aesop's famous fable.

By comparison, younger brother Johann became the fabled hare, a mathematical prodigy both clever and quick-witted. In 1691, he traveled all the way to Paris to tutor French mathematicians in the calculus, including no less a person than the Marquis Guillaume de l'Hospital, France's most gifted man of numbers.

In exchange for 300 pounds "give me at intervals some hours of your time to work on what I request," the marquis had proposed, and also "communicate to me your discoveries . . . [but] not to any others." That last request had worried Johann, but he had consented to it nonetheless; he wanted the money and, besides, the marquis appeared to be an honorable man.

While in France, Johann became quite the disciple of the new mathematics, emerging as its most articulate and arrogant defender against growing accusations that England's Isaac Newton had been the first to invent the calculus, not Germany's Leibniz; after all, the zealous young man inveighed, Newton had published his version of the calculus in 1687, three years *after* Leibniz. (See "Apples and Oranges.")

"When in England war was declared against Monsieur Leibniz for the honor of the first invention of the new calculus of the infinitely small," an immodest Johann later would recall, "I alone like the famous Horatio Cocles kept at bay at the bridge the entire English army." The dispute would drag on for years, but it had the immediate effect of boosting Johann's career, and ego, inasmuch as people everywhere began to speak his name in the same breath as Leibniz's.

Publicly, Jakob reveled in Johann's increasing fame abroad, reminding everyone back in Basel that he had been his younger brother's mentor. Privately, though, Jakob became increasingly jealous of Johann's friendship with Leibniz and began to worry that his swaggering younger brother was proving to be a better mathematician than he.

In 1695, in order to stay closer to home, the free-roaming Johann applied for a professorship at the University of Basel. Under normal circumstances, the worldly wunderkind would have been a shoo-in; but behind the scenes, Jakob traded on his friendship with members of the university's academic senate, getting them to reject the application.

When Johann learned of his brother's betrayal, he was heartbroken and furious. But his revenge came swiftly and sweetly, when Christiaan Huygens, the Netherlands' greatest living scientist, invited him to become chairman of the mathematics department at Groningen University.

From then on, the relationship between the Bernoulli brothers deteriorated rapidly; they took to belittling each other's mathematical prowess, first privately, then publicly, in the pages of the prestigious journal *Acta Eruditorium*. (Ironically enough, it meant "The works of the erudite.") The brothers taunted each other in print for four consecutive years, until finally, in 1699, the journal's disgusted editor put an end to it.

The bitter warfare continued, however, with insults being promulgated via letters to colleagues and handbills circulated throughout Europe. Thus, while the rest of the world came together to celebrate the end of the seventeenth century, Jakob and Johann found themselves as far apart filially as their universities were geographically.

It would have been hard for anyone to believe that the warring Bernoulli brothers were devoted family men, but they were. Forty-five-year-old Jakob was married and had two children. A doting father and husband, thirty-two-year-old Johann also had two children, and his wife was about to give birth to another.

It happened less than a month into the new year, on the twenty-ninth of January: Johann and his wife became the parents of a brand new son, whom they named Daniel. Though it was too soon for even a Huguenot to predict, history was about to repeat itself; whether by destiny or by chance, this newborn was to follow in the footsteps of his father and uncle in more ways than one.

Five years after Daniel's birth, Johann decided reluctantly to return to Basel, to be near his father-in-law. The old man was ailing and claimed that being reunited with his daughter after these ten long years was the only thing that would cure him. "For we cannot answer in eternity for our stubbornness against God," a resigned Johann had responded, "if we sin against our parents by hastening their death."

En route to Basel, news reached Johann that Jakob had died of tuberculosis. Although it was a horrible way for the brothers' rancorous relationship to end, the hardened Johann saw it as a way of scoring one final victory. "This unexpected news bowled me over," he later would recall shamelessly, "and then it entered my thoughts immediately . . . that I could succeed to my brother's position."

Promptly upon his arrival in Basel, Johann began lobbying for Jakob's vacant professorship, and in less than two months he got his way. When he moved into his brother's old office, he felt a bit remorseful, but mostly he felt justified at having finally obtained the university post his older brother had been instrumental in denying him a decade earlier.

His family's unexpected move to Basel left young Daniel with only a few precious memories of the Netherlands, chief among them being windmills. Much of the country lay below sea level, so the Dutch used windmills everywhere to pump water away from their cities and farms.

Daniel also treasured the memorable images of diamond-shaped kites. They were a relatively recent invention, but many Dutch children had quickly discovered how easily and gaily the kites flew upon the strong winds blowing in from the North Sea.

Life in Basel was very different, young Bernoulli discovered, but in one respect things remained the same. His father did not like being contradicted, so he always was careful not to appear disobedient or disrespectful.

He did not disagree or complain, for instance, the day his father announced that he, Daniel, was destined to marry the daughter of a certain wealthy businessman. Neither did he object when his father decreed that he was to become a merchant; ironically, the father—like *his* father before him—wished his young son to prosper by becoming something more than an impecunious mathematician.

Secretly, however, Daniel Bernoulli was not so tractable as he appeared. For one thing, he was not at all sure he would wed that particular girl; he would decide that for himself when the time came. And furthermore—as if it were in the Bernoullis' blood—he was fascinated by mathematics and coaxed his older brother Nikolaus II into tutoring him.

Daniel Bernoulli's eyes widened as he learned about the calculus. He was equally fascinated to discover the various ways it already had been used by his father and uncle to explain subtle questions about the everyday world, such as: "What shape did a ski slope need to have in order to produce the fastest ride down?" or "Why was the shell of a chambered nautilus shaped like a perfect spiral?" or "Why were soap bubbles always round?"

The young man also was surprised to learn that Isaac Newton, whom his father maligned so tirelessly, recently had discovered the simple rules by which solids moved—something all scientists before him had failed to do, despite 2,000 years of trying. Newton's was a heroic achievement that stirred something deep within Bernoulli's mind and spirit.

Because he had assimilated during his formative years in the Netherlands some of that country's famous preoccupation with water, Daniel Bernoulli now wondered whether Newton's laws could be applied to fluids. Intuitively he doubted it—fluids obviously were so different from solids—but intellectually he was not nearly sophisticated enough to decide the question—at least, not yet.

While Daniel's study of Newton continued in secret, his father's unrelenting public denigration of the vaunted Englishman escalated. Soon, in fact, it reached a point where the elder Bernoulli needed help in manning the various fronts of battle.

At first, he recruited the help of Daniel's cousin Nikolaus I and older brother Nikolaus II, but then he enjoined help from Daniel himself. The youngster demurred, pretending to be disinterested in his father's war of words, though in truth, he had come to admire Newton and hoped one day to be just like him.

It was the first time the thirteen-year-old had defied his father openly. Instead of being angry, however, the despotic Professor

Bernoulli was only mildly irritated and to some extent reassured that this son of his was definitely *not* destined to become a mathematician.

Later that year, however, it became equally clear that his son was not fated to become a merchant. After twice trying to make a go of being an apprentice in the pharmaceutical business, young Bernoulli ended up failing as completely as his father had failed a generation earlier.

After that, Daniel Bernoulli decided to stop altogether his pretending to go along with his father's astrologylike notions of what God expected of him—that included the business of becoming a merchant, marrying some preselected girlfriend, *and* the nonmathematical charade he had been carrying on for several years now. Consequently, the young man broke the bad news to his father and begged for permission to pursue his love of numbers.

This time round, his stern father's reaction was truer to form. The young man *could* continue his mathematical studies, the senior Bernoulli snarled angrily, but being a *professional* mathematician was absolutely out of the question; instead, he decreed, his son would become a doctor.

The only part of that scolding which the young man heard clearly was his father giving him permission to pursue his mathematical interests. As for the rest of it, he would obey his father, albeit halfheartedly, because he saw no harm, and possible advantages, in acquiring medical training.

For the next several years, Daniel Bernoulli attended the university and was tutored at home by his kindly and patient older brother Nikolaus II. This only strengthened their relationship, which already was as warm and caring as that which once had existed between two other brothers, their father and uncle.

In time, as it became obvious that young Bernoulli's interest in mathematics was no mere fleeting fancy, his unhappy father relented and offered to tutor him. It was a rare honor to be taught the calculus by the very man whom Leibniz considered his closest friend; unfortunately, it was also a rare punishment.

One day, for example, the merciless professor gave his son an exceptionally difficult problem to solve. After struggling for hours, the young man finally solved it, whereupon he walked to his father's study and handed in his work.

Proud of his achievement, the youngster eagerly awaited his father's praise. The work had been done correctly, the senior Bernoulli groused, but "couldn't you have solved it right away?"

Insensitive though he was, Professor Bernoulli shared generously with his son everything he knew about mathematics and natural philosophy. During one lesson, for example, he began describing an exciting new idea that was to prove crucial in the youngster's career; it concerned energy, though it had not yet been given that name.

Instead, like his illustrious friend Leibniz, the elder Bernoulli called it *vis viva*—Latin for "living force"—because it appeared to be something possessed by objects that were to some degree animated. By doing various experiments, Leibniz had noticed that an object's *vis viva* depended on only two things: its mass and speed. Mathematically speaking, if m stood for an object's mass and v its speed, then the formula for *vis viva* boiled down to this:

$$VIS\ VIVA = m \times v^2$$

A rogue elephant, being massive and fast, had a lot of *vis viva*. A leaf blown along by a gentle breeze, being lightweight and slow, had very little *vis viva*. A seated young Daniel Bernoulli, listening raptly to his father's lectures, had no *vis viva* whatsoever.

Vis viva was like some kind of invisible fuel, the young man was told; it could be spent in order to raise an object off the ground. *Vis viva*, for example, was what propelled a rubber ball tossed into the air; as its altitude increased, its *vis viva* decreased.

At the top of its climb, exhausted of all its *vis viva,* the ball stopped and began to fall back down again. Along the way, experiments indicated, the ball recovered fully all of its spent *vis viva*—which was like some perfectly recyclable fuel—so that when the ball returned to its starting point, things were back exactly where they had started.

Throughout the ups and downs of a thrown ball's existence, in other words, there was a precise give-and-take between altitude and *vis viva.* When one increased, the other decreased, so that the total of the two never changed:

$$\text{ALTITUDE} + VIS\ VIVA = \text{CONSTANT}$$

It was as if an object's *vis viva* could never be destroyed, merely exchanged for something else—in this case altitude. At least, that was what Johann Bernoulli, Leibniz, and many others firmly believed; they called it the "Law of *Vis Viva* Conservation." (Late in the next century, scientists would call it the "Law of Energy Conservation" a sacred tenet of modern physics.) (See "An Unprofitable Experience.")

Though many of these lessons were intellectually daunting, young Bernoulli learned them well. He was a genuine prodigy, graduating from college when he was but fifteen years old. A year later, in 1716, he earned his master's degree and immediately began his medical schooling.

Given his training, it was natural for the teenage medical student to think of the human body as merely a complex machine, like a fancy watch, subject to the elucidation of scientific laws. According to that mechanistic way of seeing things, the body was not animated by some supernatural soul, as Aristotle and many after him had believed, it was powered by *vis viva;* its overall movements, furthermore, conformed to Newton's laws, just as with any other solid object.

As he pursued his medical studies, young Bernoulli was delighted to discover that others shared his nuts-and-bolts philosophy. In his book *On the Motion of Animals,* for example, Giovanni Alfonso Borelli shot down one of humanity's most wistful and wonderful dreams. After computing the vastly disparate abilities of human and bird muscles to store up *vis viva,* he concluded: "It is impossible that men should ever fly craftily by their own strength."

Young Bernoulli found a kindred spirit also in British physician William Harvey. Up until now, nearly everyone had followed Aristotle, Hippocrates, and Galen in believing that the heart was the human body's primary source of *heat.* But in his book *On the Movement of Heart and Blood in Animals,* Harvey had written the heart was like a pump and our blood vessels like a network of canals: "The one action of the heart is the transmission of the blood and its distribution, by means of the arteries, to the very extremities of the body."

Young Bernoulli was attracted to Harvey's research, because it suggested a way he could indulge two of his loves, mathematics and fluids, while earning the medical degree his father expected of him. Furthermore, it was a challenge worthy of his best efforts, considering that no one—not Newton, Leibniz, or even the imperious Johann Bernoulli—had yet discovered the basic rules by which fluids moved.

Indeed, young Bernoulli's father was currently embroiled in an argument over Newton's analysis of water streaming from a hole punched near the bottom of a drinking cup. Natural philosophers at the time were still very inept at measuring the speed, pressure, or even the size of fluid streams, and such uncertainties inevitably led to endless quibbling.

As always, Daniel Bernoulli remained uninvolved in his father's battle with Newton, but though he maintained his distance, he was very definitely interested in the outcome. That was because his doctoral dissertation concerned the mechanics of human respiration, which, like the water-cup problem, involved the movement of a fluid, namely air.

In 1721, following the completion of his medical studies, young Bernoulli came away sobered by the unresolved complexities of fluid behavior. Now, more than ever, he wanted to tackle the subject that had defeated so many before him; all he needed was an academic position that would give him the freedom and equipment to fulfill his dream.

Like his father had before him, the twenty-one-year-old applied for a professorship at the University of Basel. And like his father had been before him, amazingly, he was denied the opportunity, though for different reasons.

It was the practice in Bernoulli's day for a university to rely on chance in selecting from among equally qualified candidates. Finalists competing for a faculty position, therefore, were required to draw lots; the winner was awarded the professorship.

Because of his precocity, Daniel Bernoulli had been selected as a finalist for *two* professorships, one in anatomy and botany and the other in logic. Consequently, in his own mind, the brilliant young medical graduate had been rather relaxed about his chances of winning a coveted place in his hometown university; he never would have guessed that he would lose out on both lotteries.

Like his father, Bernoulli had grown up believing in the ability of the calculus to predict the outcome of all rational processes. Now, however, the twenty-one-year-old had discovered that, the calculus notwithstanding, it was still beyond the ken of science to predict the outcome of a game of chance, let alone a person's life.

VIDI

Even though we humans have always lived on solid land, we owe our existence to fluids. Without water to drink, we would die in a matter of days; worse still, without air to breathe, we would perish in a matter of *minutes.*

Fortunately, the earth is awash with water and air; indeed, there has always been plenty of both to sustain our 4-million-or-so-year-old species. Unfortunately, though, we have not always been as adept as we are today at exploiting those precious resources.

Our nomadic, cave-dwelling ancestors, for example, were at the mercy of their region's geology. Air was plentiful everywhere they roamed—save for the tops of very tall mountains—but individuals would live or die depending on their ability to locate natural sources of potable water along their migration routes.

As our predecessors organized into cities, they settled near rivers and began to contrive ways of channeling the constantly flowing waters into their homes and onto their crops. "Egypt," Herodotus once wrote, "was a gift from the Nile."

Five thousand years ago, engineers began to build dams, canals, and aqueducts to domesticate the wild waters of the earth's great rivers, but in so doing, they relied solely on intuition or hit-and-miss experiences. As recently as 2,000 years ago, not even Aristotle had figured out the scientific rules that described how water moved.

By comparison, solid objects were simpler to study than water, because at least they held themselves together. If a rock was hit by a paddle, for example, all its parts moved in unison, making it relatively easy to describe its trajectory.

However, when struck by that same paddle, water splattered every which way, becoming a shower of countless droplets. The incohesiveness of that life-sustaining fluid, in the minds of many, made it appear incoherent.

About the only sensible thing Aristotle was able to deduce about fluids concerned their density—or, actually, the *opposite* of density, which he referred to as subtlety. "If air is twice as subtile as water," he wrote, "[a] body will require half the time to traverse the same distance in air as in water." That is, it's twice as easy to move through air than through water.

Aristotle's assertion was a reasonable guess, but eventually it was proven wrong: The resistance of an airplane fuselage moving through air, let us say, was *not* half the resistance of a similarly shaped submarine hull moving through water. The connection between a fluid's density and its resistance to moving objects turned out to be far more complicated than that.

The first person to discern a correct (and surprising) truth about the quirky behavior of water was the Sicilian scholar Archimedes. It all began when his friend Hiero II, monarch of Syracuse, wondered whether there was any way of checking the purity of the metal that had been used to make his newly wrought imperial wreath. It was supposed to have been made of pure gold, but the king suspected the royal goldsmith of having diluted it with silver.

It was a bedeviling problem, because the wreath could not be scraped for a sample nor defaced in any other way. Archimedes racked his brain day and night, but without success. Then, one afternoon, he decided to take a dip at the public baths.

He had done this often, as a way of giving his mind a rest. A thousand times, in fact, he had lowered his sizable body into a pool of water and not paid any attention to how, invariably, the water level went up a fraction of an inch; this time was different.

Archimedes was so excited by his revelation that he ran home before getting dressed, shouting like some naked lunatic "I have found it, I have found it!" What he had found, he later revealed in a book called *On Floating Bodies,* was the Law of Buoyancy, according to which a floating object always pushed aside an amount of water equal to the object's own *weight.*

Archimedes also had discovered that a nonfloating object—one that sank—pushed aside an amount of water equal to the object's own *volume.* (Today's cooks use this principle when they plunge a spoonful of shortening into a graduated cup of water in order to measure its volume.) Quite by accident, Archimedes had found a way to help his royal friend.

By placing into a tub of water Hiero's heavy metal wreath— which was nonbuoyant—Archimedes invoked his newfound revelation to determine its volume. Next, he *weighed* the wreath, whereupon he was able to ascertain its density simply by dividing its weight by its volume.

It came out to be somewhere between 10.5 and 19.3 grams per cubic centimeter, the densities of silver and gold, respectively. Archimedes had confirmed the king's suspicions that the wreath had not been made out of *pure* gold; as a result, the royal goldsmith was executed.

Though Archimedes's discoveries were important, they pertained solely to fluids in a container of some kind, such as a tub of water. Consequently, the Archimedean laws represented the beginnings of *hydrostatics,* the study of captive fluids; they had nothing to say about freely flowing water, which remained an ineluctable mystery.

Throughout the many centuries of the Roman Empire, nevertheless, engineers managed to build impressive public aqueducts that delivered as much freely flowing water per capita as many modern cities receive today. In A.D. 97, Rome's great water commissioner Sextus Julius Frontinus boasted: "Will anybody compare the idle pyramids or those other useless though much renowned works of the Greeks with these aqueducts?"

The remarkable water works were made possible not with brain but brawn. Whatever scientific principles Frontinus and his fellow engineers invoked were rather trite, such as: "Water always flows downhill, never uphill" or "The largest amount of water a pipe can deliver depends on the size of its opening." (They did not even take into account that it also depended on the water's *speed:* A pipe's output was greater, naturally, if water poured out of it faster.)

It took *fourteen* more centuries for another Italian, Leonardo da Vinci, to make the first significant discovery about moving water. Indeed, another two centuries hence, his prescient observations were to play a pivotal role in Daniel Bernoulli's own historic discovery concerning moving fluids.

For long stretches of time, the great Renaissance painter-philosopher-engineer would sit near waterfalls and toss grass seeds at them. As he watched the seeds get caught up in the water's roiling plunge downward, he sketched their paths, thus becoming the first person to illustrate in extraordinary detail the many hitherto-invisible subtleties of water in motion.

The more waterfalls he sketched this way, the more he began to realize something very important about water: Seemingly chaotic though its motion appeared to the casual observer, there were predictable patterns to its behavior, clearly revealed in the gently curving lines of the sixteenth-century master's meticulously penned drawings.

Leonardo also studied rivers, tossing seeds or sawdust into their waters and watching what happened. It was in this endeavor that he made his most historic observation, though he did not see it all at once; it dawned on him in stages.

At first, Leonardo noticed simply that: "A river of uniform depth will have a more rapid flow at the narrower section than at the wider." In other words: A river of water always flowed fastest when squeezing through a bottleneck (a potentially dangerous fact that any white-water rafter understands instinctively).

Leonardo went one step further, observing that the water's

speed increased in *direct proportion* to the narrowing of the bottle-neck. For example, through a bottleneck *one-half* as wide as the normal river, the water passed *twice* as fast as normal. Through a bottleneck *one-third* as wide, the water moved *three* times as fast, and so forth.

Leonardo's all-important discovery of so simple a fact of Nature came to be called the "Law of Continuity." Though it referred to a fluid, the law's implications could be understood by imagining, let us say, a steady stream of animals flowing into Noah's Ark.

Suppose that in this imaginary ark, pairs of animals walked through its front entrance shoulder to shoulder. Suppose also that after being processed by Noah in a giant antechamber, they filed singly through a narrow interior doorway, the ark's bottleneck, into the rear holding pens; the interior door, say, was *one-half* as wide as the front door.

According to Leonardo's Law of Continuity, in order to keep things moving along, each animal needed to *double* its speed when squeezing through the ark's bottleneck. Suppose, for example, that animals sauntered into the Ark at one pair per second—that is, two individuals every second. When each pair split up in order to file into the rear pens one at a time, each individual needed to speed up, needed to zip through the bottleneck in one-half second—twice as fast; otherwise the orderly procession of animals would back up.

In time, Leonardo surmised that his Law of Continuity applied to fluids of all kinds, including air. In fact, he was the first person in history to appreciate fully that air and water were birds of a feather. "In all cases," he wrote, "water has great conformity with air."

Leonardo had noticed this kinship as a result of studying the flight of birds in air and of fish in water. He was inspired by the former to sketch fanciful designs of man-powered ornithopters and by the latter to sketch designs of underwater ships, foreshadowings of the airplane and submarine, respectively.

In the decades immediately following these unprecedented insights, several natural philosophers made various other important discoveries about moving fluids, though none turned out to be as crucial as Leonardo's. For some reason, furthermore, all the researchers—every single one of them—was Italian!

Perhaps it was because of the tradition that had been inspired by the famous Roman water works. Or perhaps it was because of the incomparable Italian tradition during the Renaissance of world-class creativity and scholarship. Whatever it was, during the entire seventeenth century, Italians—including Galileo Galilei, Evangelista Torricelli, and Domenico Guglielmini—studied fluids with more success than any other people on earth.

Things changed dramatically in 1642, however, when Galileo died, after having been arrested by the Catholic Inquisition and forced to recant certain of his scientific beliefs. (See "Apples and Oranges.") After that, the river of ideas and inventions that had made Italy such a creative Mecca never again flowed as freely.

During those tumultuous years, science sought fertile ground elsewhere. It found it in Germany, England, France, and virtually any other country where the Catholic orthodoxy did not hold sway. This was the beginning of a new era, the climactic stages of a scientific revolution that was being aided by the religious revolution that Martin Luther and John Calvin had begun more than a century before.

Already, with their emphasis on hard work, the Calvinists were being credited with leading the seventeenth-century rise in capitalism. Now, with their emphasis on mental discipline, they were being credited with supporting the rise of scientism, the belief that ultimately everything in the natural world could be explained mathematically and proven experimentally.

In Germany, Gottfried Wilhelm Leibniz became one of the world's leading expositors of this philosophy-*cum*-religion, along with England's Isaac Newton and Switzerland's most illustrious family of Huguenots, the Bernoullis. They would not succeed entirely, but in the decades ahead, these natural philosophers were to take on three of the most perplexing mysteries of Nature: first solids, then fluids, and, finally, human beings themselves.

VICI

In 1723, Daniel Bernoulli ran away from Basel in order to forget his having failed to win a university professorship. The crestfallen young doctor headed for Italy, hoping to practice medicine there, but when he arrived in Padua, he himself became deathly ill with a fever.

During his year-long recuperation, Bernoulli corresponded with a friend named Christian Goldbach, in the course of which he restated many of the lessons he had learned from his father. Moreover, he applied those techniques to many of the most provocative problems of his day, including the one about how water gushed forth from a hole in a drinking cup.

Once recovered and raring to flex his intellectual muscles, Bernoulli decided to enter the annual competition sponsored by the French Academy of Sciences. This year's challenge was to design a ship's hourglass that would produce a reliable trickle of sand or water even when tossed from side to side by rough seas.

Far from being academic, the problem was of crucial importance to sailors, who relied on clocks to compute their longitude—that is, their easterly/westerly distance from home port. (Latitude was reckoned easily enough by observing the sun's position.) For that reason, countries were competing fiercely to come up with accurate onboard chronometers, knowing that with superior navigation came superior profits from improved maritime trade.

Young Bernoulli submitted his entry but really did not expect to win. Recently having lost out on two lotteries, he was not feeling very lucky; and besides, he had come to discover that the world abroad was full of highly talented mathematicians, many of whom were competing against him for this award.

When the results were announced, therefore, twenty-four-year-old Bernoulli was flabbergasted to learn that he had won first prize! His award-winning design had involved mounting an hourglass atop an iron slab floating in a pool of mercury; even when buffeted by violent storms, the young man had calculated, the sheer heaviness of the mercury would keep the timepiece from sloshing around very much, providing it with a relatively stable foundation.

Bernoulli had barely recovered from the surprise of winning the French Academy's inestimable award when he received still more shocking news. Goldbach had been so impressed with the letters he had received from the convalescing Bernoulli, he had decided to have them published.

Though Bernoulli objected, complaining the letters had been written informally, without proper attention to detail, he ended up relenting, giving the book his blessing and its unprepossessing title: *Some Mathematical Exercises.* Furthermore, out of respect for his father, whose ideas had inspired much of what was contained in the missives, the unassuming young man asked the publisher to identify him simply as "Daniel Bernoulli, Son of Johann."

In 1725, having gone from being a double loser to a double winner, a rejuvenated Bernoulli decided he had seen enough of Italy and headed home. When he arrived in Basel, however, his homecoming was nothing like he had expected it would be.

In recent months, letters had poured in from all over the world, hailing Bernoulli's new book as the work of a mathematical prodigy. Most amazingly, awaiting him was a letter from Catherine I, the Empress of Russia.

In the note, she praised the young man's uncommon talents and invited him to become professor of mathematics at the Imperial Academy of Sciences in St. Petersburg. Both city and Academy having only recently been built, by order of her just-deceased husband Peter the Great, the empress was now attempting to populate them with the finest minds in all of Europe.

Bernoulli was flattered by the offer but intimidated by the prospect of being alone so far from home. He was tired of living abroad; he craved sleeping in his own bed and being close to his own family.

He decided to turn down this opportunity of a lifetime and began to compose a letter of regret to the empress. But before young Bernoulli could put pen to paper, his older brother Nikolaus II interceded, selflessly offering to go with him.

With that, an emboldened Daniel Bernoulli decided to accept the empress's offer, on condition that she grant professorships to both him and Nikolaus II. "If you could second this plan," he explained to an Academy official, "you would have the merit of keeping together two brothers bound by the closest friendship in the world."

With the empress's whole-hearted approval, both brothers left Basel in the fall of 1725, journeying across Europe on the longest trip of their lives. About two months later, they arrived in St. Petersburg and almost immediately regretted having gone there.

The Russian people themselves were warmhearted and friendly, but their weather was cold and nasty. At the beginning of the new year, Nikolaus II came down with a respiratory infection that wouldn't go away. It persisted into the spring and summer, until finally, on July 26, 1726, he succumbed to the ravages of tuberculosis.

Jolted by this cruel twist of fate, Daniel's impulse was to go right back home and put the traumatizing incident behind him. But his belief in destiny made him decide to stick it out in St. Petersburg. There had to be a reason other than to see his brother die, the young Huguenot consoled himself, to explain why God had brought him to this faraway place.

In an attempt to assuage his loneliness, Daniel Bernoulli decided to send for Leonhard Euler, a young man whose intelligence had earned rave reviews from the severe Professor Johann Bernoulli. Indeed, Daniel could not remember a single instance when his father had ever complimented anyone's sagacity so uninhibitedly, save his own or Leibniz's.

Leonhard Euler's ancestors, like Bernoulli's, had originally fled to Basel in order to escape religious persecution, and they had prospered. The only difference was that young Euler had descended from a long line of comb makers and clergymen, not spice merchants and pharmacists.

Euler himself was the son of a Calvinist minister in a small town just down the river Rhine from Basel. Before he was born, his father had often trekked to the University of Basel to hear Professor Jakob Bernoulli lecture on the mathematics of everything from astronomy to zoology.

After Leonhard was born, the minister had passed on to him everything he had learned from those lectures. Subsequently, the Reverend Euler had recognized in his young son a genuine talent for numbers and arranged to have him attend the best schools in Basel.

In 1720, at the tender age of thirteen, young Euler had matriculated to the University of Basel. Soon afterward, the boy genius had asked to be coached by the famous Professor Johann Bernoulli; but "he was very busy," Euler had lamented, "and so refused flatly to give me private lessons."

The elder Bernoulli had, however, deigned to allow the gifted youngster to drop by every Saturday afternoon to have his work evaluated. During those encounters, Euler had solved every single problem put to him by the surly professor, and in record time. Eventually, therefore, the brilliant teenager had been granted the extreme privilege of becoming Bernoulli's premier protégé.

Back in 1725, just weeks before Daniel and his brother had left for St. Petersburg, their father had surprised them by expressing increasing admiration for the wonder child. Indeed, Professor Bernoulli had come to sound like a man unabashedly in awe of a scientific and mathematical mastermind the likes of which the world rarely beheld.

With a recommendation such as that, both Daniel and his brother had entreated Catherine I to consider inviting the adolescent wizard to her young Academy. Now, with his brother dead and Euler's studies completed, Daniel Bernoulli pressed even harder, and successfully, to have Euler invited there.

While awaiting Euler's reply to the empress's invitation, the twenty-six-year-old Bernoulli tried to pick up where he had left off in his study of the human body. Having grappled with the problem of respiration, he now turned his attention to the even more complicated problem of *blood* circulation.

Most of what was known in his time about the insides of the human body was the result of vivisections that had been conducted on and off for more than 2,000 years. It was a macabre practice that had been described and defended in ancient times by the Roman scholar Celcus in *On Medicine,* the first volume of his epic encyclopedia:

> Thus, they laid open men whilst alive—criminals received out of prison from the kings—and whilst these were still breathing, observed parts which beforehand nature had concealed . . . Nor is it, as most people say, cruel that in the execution of criminals, and but a few of them, we should seek remedies for innocent people of all future ages.

During the Renaissance and throughout Daniel Bernoulli's day, human vivisections were still done, though the complexity of the body's insides often left scientists more bewildered than ever. "When I first gave my mind to vivesections," William Harvey had complained, "I found the task so truly arduous . . . that I was tempted to think . . . that the motion of the heart was only to be comprehended by God."

Nevertheless, through sheer perseverance, science came to discover that the body's insides were infiltrated by veins and arteries of various diameters, some wide, some narrow. Moreover, by watching the arteries of people who were still alive, Harvey and others had confirmed that when the heart contracted, the arteries were suddenly and momentarily filled with blood, causing them to bulge, like so many over-stuffed sausages.

Harvey and his contemporaries also had found that when the heart relaxed, the arterial walls snapped back, squeezing down on the blood within them and squirting it on its way. Over and over again, the arteries bulged and squeezed, bulged and squeezed, producing what philosophers long ago had called the "pulse" of life.

What Daniel and others of his time wished to know was the *speed* and *pressure* with which blood actually flowed through our complex circulatory system. It was the kind of problem the brilliant Roman water-works engineer Frontinus might have pursued, but had not.

No one throughout the centuries following Frontinus had done so either, simply because the problem was so complex. "Those who have spoken about the pressure of water flowing through aqueducts," Bernoulli complained, "did not hand down any laws other than those for extended fluids with no motion"—that is, the hydrostatics founded by Archimedes.

In the case of *static* fluids, philosophers had no difficulty computing pressure; they simply divided a fluid's weight by the area of its supported surface. It was a straightforward adaptation of the definition that always had been used to calculate the pressure of *solids*.

For example, the solid point of a high-heel shoe worn by an average-size woman produced huge pressures—about 2,000 pounds per square inch!—because her entire weight was being supported by a very small area. (In fact, in the early days of air travel, passengers wearing spiked heels were prohibited from boarding an airplane, owing to the danger of their puncturing its thin metal floors.)

Similarly, the static waters trapped within an artificial reservoir produced uncanny pressures on a dam. Why? Because the water's voluminous weight was being supported—kept from spilling out—by the relatively small surface area of the dam's wall. (In the case of Hoover Dam, in Nevada, waters press against the concrete wall with pressures up to 45,000 pounds per square foot!)

By contrast, for *freely moving* fluids, the situation was far more complicated. That was because it was not so easy to measure, or even define, the pressure of something whose weight was constantly shifting or whose shape—and, therefore, the area of whose supported surface—was constantly changing.

For Daniel Bernoulli's generation, this was more than a theoretical problem. Many an eighteenth-century physician was in the habit of treating patients by deliberately cutting open one of their veins, the belief being that people became swollen with illness, because their bodies had accumulated a surfeit of blood. Known as phlebotomy, or bloodletting, the procedure dated back to the fifth century B.C., when Hippocrates used it on patients with inflammatory ailments. In Bernoulli's time, however, many physicians used the technique to treat nearly *any* kind of illness.

The practice had become so popular, so gratuitous, in fact, there arose a demand for ways of finessing its brutalizing effect on patients. If someone could invent a way of measuring a patient's blood pressure, then physicians could use that information to gauge exactly how extensively they should bleed him.

The question was "How could such a thing be done?" No device existed to measure blood pressure; incredibly, no reliable gauge of *any* kind existed to measure the speed and pressure of *any* kind of fluid moving within *any* kind of pipe-shaped conduit.

In 1727, while young Bernoulli pondered this matter, news

came of Isaac Newton's death. The great natural philosopher had been a font of creativity and always would be remembered for discovering, among other things, the three truisms concerning the behavior of solid objects:

Truism #1: A solid object will move in a straight line at a constant speed (or not move at all), unless it is pushed off-course by some force.

Truism #2: A solid object will invariably accelerate (or decelerate) if it is pushed by some force.

Truism #3: Two solid objects pushing off of each other will feel equal and opposite forces.

In Basel, Daniel Bernoulli's father reacted ambivalently to the demise of his old nemesis, the darling of "the scurvy English." On the one hand, it pleased him to know that now he alone would be the most esteemed mathematician in the whole of the civilized world; but on the other hand, Newton's death brought to mind his own mortality.

In reflecting back on his stormy career, the sixty-year-old professor felt downright cheated by life. For example, the Marquis de l'Hospital—the French mathematician whom many years ago Bernoulli had tutored in the calculus and in whom he had confided many of his discoveries—had turned out to be something of a scoundrel. Not only had the marquis taken credit for some of those discoveries and failed to pay Bernoulli for past services, he recently had written a best-selling calculus textbook without including a proper acknowledgment to his former mentor.

And then there was the matter of his wife and in-laws: In order to please them by staying close to home, he had constantly declined very desirable offers from prestigious universities around the world. Consequently, he had been stuck in the same post at the same provincial university all of his adult life.

Then there was the most grievous insult of all, the one that angered him most: After all these many years of trying, the great and glorious Professor Johann Bernoulli had failed to win first prize—or even an honorable mention!—in the French Academy's world-famous competition. Even his own young son was ahead of him on that score.

Thousands of miles away, in St. Petersburg, Newton's death had a far different effect on Daniel Bernoulli. The young man had never met Newton but felt an emotional attachment to him that came partly from wishing one day to be as famous as he. "Newton, a man immortal for his merits," Bernoulli eulogized, was "superior and incomparable in his abilities."

Two people close to his heart having died in as many years, young Bernoulli was delighted when finally the day came that Euler arrived at the Russian Academy. Also, he was thrilled to learn that his father's nineteen-year-old prize pupil had just won a prestigious Certificate of Merit in the French Academy's annual competition.

Cheered by Euler's stimulating intelligence and youthful energy, Bernoulli soon began to regard the St. Petersburg Academy with new appreciation, which it fully deserved. During its few years of existence, the young and prestigious institution had attracted natural philosophy's *crème de la crème* and provided them with the very best facilities.

"I and all others who had the good fortune to be for some time with the Russian Imperial Academy," Euler would recount one day, "cannot but acknowledge that we owe everything . . . to the favorable conditions we had there."

In the years to come, Bernoulli and Euler would labor on many of the same problems, separately and collaboratively. They both would make historic discoveries concerning solids and fluids, but whereas Euler proved to be more of a pure mathematician, preferring to work in his office with paper and quill, Daniel had proved early on that he was not above getting his hands wet in a laboratory.

Shortly after Euler's arrival, in fact, Bernoulli resumed his efforts to find a way of measuring the pressure of water moving through a pipe. He tinkered with polished iron pipes of various diameters but failed continually in his purpose.

Nearly fifty years earlier, a clever Frenchman named Edme Mariotte had managed to measure the pressure of water not flowing *through* a pipe but gushing *out* of one. He had done so by letting the escaping water push against one end of a small wooden teeter-totter. At the other end of the teeter-totter, Mariotte had placed lead shot. By the amount of weight it took to balance the push of the water, Mariotte had been able to estimate its force and, from that, its pressure.

Certainly, it would not be wise to use Mariotte's technique to measure blood pressure; that would require cutting open a person's artery and letting the blood gush out in huge quantities. For Bernoulli, therefore, the trick was to come up with a way of measuring a fluid's pressure *without* hemorrhaging the fluid or noticeably disrupting its flow through the pipe.

In 1729, as Bernoulli contemplated the matter, he remembered something he had read in Harvey's book. "When an artery is divided or punctured," the vivisectionist had noted, "the blood will be seen spurting with violence." During the course of a complete heartbeat, Harvey continued, the jet of blood "projected now to a greater, now to a less distance," the tallest jet occurring "when the heart contracts."

Clearly, Bernoulli reasoned, the *height* of the spurting blood was a direct measure of its *pressure* within the artery; the greater the arterial pressure, the higher the spurt. As our heart contracted and relaxed, our blood pressure increased and decreased, the highs and lows corresponding to what doctors called the *systolic* and *diastolic* pressures, respectively.

Following Harvey's lead, Bernoulli punctured the wall of a pipe and attached to this small hole one end of a glass straw. Allowing water to flow through the pipe as usual, he watched, waited, and then noted with elation that as the water flowed past the opening, a small column of water rose up in the glass tube and stopped at a certain height. He had done it! That height was a measure of the flowing water's pressure.

If the water rose high up the glass tube, it meant that, at that point, the water pressure within the iron pipe was large. Conversely, if the water barely rose up the glass tube, it meant that, at that point, the water pressure within the iron pipe was small. And in all cases, happily, no water was spilled in making the measurements.

Eager to share the news of his breakthrough with others, Bernoulli wrote to his old friend Christian Goldbach, who was now in Moscow. "During these last days, I made a new discovery that will be very useful in the design of the water supply," an elated Bernoulli predicted, "but mainly, it will open a new era in physiology."

True to Bernoulli's prophesy, physicians all over Europe soon began to adapt his innovation to their work. Before deciding to cut open a patient's vein to bleed him, doctors now stuck pointed-end glass tubes directly into one of his arteries.

The blood continued to flow largely unspilled and uninterrupted through the punctured artery, but a tiny bit would rise up the glass tube. Where it stopped, high or low, was invariably a measure of the patient's blood pressure. (Incredibly, it wouldn't be until 1896 that Italian doctor Scipione Riva-Rocci would invent the sphygmomanometer, the painless, inflatable cufflike device familiar to us today.)

Bernoulli was thunderstruck by the implications of his new technique, not only to medicine but to the physics of fluids. "I cannot but feel well concerning those physical principles with which I became strongly involved," Bernoulli observed modestly at the time, "since indeed they led me by the hand to exposing many new properties concerning . . . the motion of fluids."

It was quite possible, he sensed with a quiet, mounting excitement, that he finally had come to the threshold of realizing his lifelong dream of becoming the Isaac Newton of this exceedingly slippery subject. But this was no time to stop and daydream about some childhood fantasy.

As he pressed onward, the young man confirmed what Leonardo da Vinci had first discovered two centuries before, the Law of Continuity: Water flowing from a wide pipe to a narrower one sped up; water flowing from a narrow pipe to a wider one slowed down.

What Bernoulli observed next, however, was completely unprecedented. Slow-moving water (in the wide pipe) always had a higher pressure, he discovered, than fast-moving water (in the narrow pipe). In other words, there appeared to be a trade-off between pressure and speed: The smaller the speed, the greater the pressure, or the greater the speed, the smaller the pressure.

Immediately Bernoulli's mind flashed on Leibniz's famous principle, the Law of *Vis Viva* Conservation. Bernoulli's father had taught him that it applied only to solids. But now, the young man wondered: Was it possible that he had stumbled on evidence that fluids, too, obeyed an analogous kind of conservation principle?

Bernoulli's heart quickened at the thought, and so did his mind. According to the conservation principle, when any object was tossed into the air, there always was a trade-off between its *vis viva*—that is, its energy of motion—and its altitude. If Bernoulli's hunch was right, his new principle would involve a trade-off between a fluid's *vis viva* and its *pressure*.

Before going any further, however, Bernoulli had to stop and think about what he was saying. Leibniz's formula for *vis viva* applied solely to solid objects:

$$VIS\ VIVA = m \times v^2$$

Was it possible to extend its meaning to include fluids, young Bernoulli wondered, and if so, how?

Ironically, finding the answers required him to call upon the mathematical ideas of both Leibniz *and* Newton. In life, the two arch rivals had never seen eye to eye about anything; now their brain children were about to collaborate in a most felicitous way.

Guided by Leibniz's calculus, Bernoulli began by breaking this complex problem down to its infinitesimal parts. In particular, he imagined slicing the water flowing through a cylindrical pipe into an infinite number of infinitely thin wafers—so thin they could not be discerned with the aid of any conceivable laboratory device.

Bernoulli imagined these impossibly thin watery wafers to behave like a bumper-to-bumper parade of *solid* rubber pucks, pushing against each other through the pipe. In effect, Bernoulli was imagining that even though fluids and solids behaved differently on a macroscopic scale, essentially they amounted to the same thing when looked at through the infinitely powerful microscope of the mathematical imagination.

Next, Bernoulli used Newton's famous three truisms of solid behavior to calculate the pushings and shovings between his hypothetical solidlike watery wafers. And finally, to obtain the net result, the young man used Leibniz's calculus to add up the infinitude of wafer-to-wafer interactions.

The normally ceremonious Bernoulli danced with glee: His calculations had led to a fluid version of Leibniz's old *vis viva* formula. In fact, both formulas were identical, except for one very understandable substitution: In place of the mass of a solid object there appeared a reference to the density of a fluid, symbolized by ρ, the Greek letter *rho*. That is:

$$VIS\ VIVA = \rho \times v^2$$

A fast-moving deluge of very dense molasses, for example, had a huge *vis viva,* a huge energy of motion. By contrast, a slow-moving trickle of very thin alcohol had very little *vis viva.* And static fluids, like the tears of joy that now pooled up in Bernoulli's eyes, had no *vis viva* at all.

What's more—and this was the most exciting part—Bernoulli's calculations had confirmed what his pipe experiments had first suggested to him: Fluids do obey their own version of the old Law of *Vis Viva* Conservation. "I thus added a new portion to the theory of water," Bernoulli enthused, "with the most pleasing success."

As in the case of the *vis viva* formula itself, Bernoulli's new fluid version of the conservation principle was nearly identical to the original solid version. The only difference was that a moving fluid traded off its *vis viva* for pressure, not altitude:

$$\textbf{PRESSURE} + \textit{VIS VIVA} = \textbf{CONSTANT}$$

In terms of mathematical symbols, using P to stand for pressure, Bernoulli's revelation boiled down to this:

$$P + \rho \times v^2 = \textbf{CONSTANT}$$

Bernoulli's discovery could be considered in terms of a lobbyist trying to persuade senators to vote her way on some political issue. The more swiftly she made her rounds—the more she divided her time—the less she was able to pressure each politician; similarly, in the case of a fluid making the rounds, the faster its speed (the greater its *vis viva*), the smaller was the pressure it exerted on its surroundings.

The same was true the other way around. The more slowly the lobbyist made her rounds, the more she was able to pressure each politician; similarly, the more slowly a fluid moved (the lesser its *vis viva*), the greater was the pressure it exerted on its surroundings.

Bernoulli's chain of reasoning applied perfectly well to blood racing through an artery. Each time the heart pumped, the artery ballooned out (widening its diameter), causing the blood flowing through it to slow down, in accordance with Leonardo's old Law of Continuity. That meant, according to Bernoulli's new principle, the blood's *vis viva* decreased and its pressure increased.

Conversely, each time the heart relaxed, the artery squeezed back down again. The blood sped up through the narrowed vessel—that is, its *vis viva* increased momentarily—and its pressure decreased accordingly. And so it was with *all* liquids, Bernoulli had discovered, moving through all kinds of channels.

A century later, a German physicist named Gustave Gaspard Coriolis would add a factor of one-half to the original formula for *vis viva*. He did this while working on a wholly different problem—one having to do with the earth's spin—solely for the convenience of his own calculations, but his version of the formula stuck. From then on, therefore, Daniel Bernoulli's principle came to be written:

$$P + \rho \times \tfrac{1}{2}\, v^2 = \text{CONSTANT}$$

In a way, this remarkable equation was not just a summary of fluid behavior but a validation of Bernoulli's mathematical career. It could be argued that the thirty-year-old had stumbled upon it by accident or been guided to it by destiny, but either way, the equation's elegant simplicity, its poetic conciseness, left no doubt that a great truth had been articulated. "It is clearly very amazing," the young author marveled in the afterglow of discovery, "that this very simple rule, which nature affects, could remain unknown up to this time."

Unable to contain himself, Bernoulli confided his discovery to a few friends at the Academy, most especially Euler, with whom he had developed a brotherly attachment. Euler himself had not been doing too badly, publishing more papers than anyone else at the Academy, on subjects ranging from astronomy to military weaponry to the movement of solid objects with complicated shapes.

As Euler's reputation had increased, so had the reverence with which he had been greeted in letters sent by his aging mentor back in Basel. Several years earlier, Euler had rated but a moderately flattering salutation—something along the lines of "esteemed colleague"—but in the latest communication, Johann Bernoulli had addressed him unrestrainedly as the "most learned and gifted man of science, Leonhard Euler."

It was well-deserved approbation from a man who himself was feeling particularly learned and gifted this year. It was 1730, and to Professor Johann Bernoulli's enormous joy and relief, he had finally won, in his words, "the great prize of 2500 Livres of the [French] Royal Academy of Sciences."

When Daniel received word of his father's tremendous accomplishment, something ached within him to return home. He had come to cherish the intellectual freedom at the Russian Academy and the pampering he had received from its royal benefactors; but his work there was done and, if truth be told, he still hated the cold climate.

For the next two years, he tried to get a position at the University of Basel, but unluckily, he kept losing the academic lotteries. In 1732, just when he was about to give up, however, the young man finally hit the jackpot, winning a coveted professorship in the departments of anatomy *and* botany.

Before leaving Russia, Bernoulli hastened to finish up one very important piece of business. During the seven years he had worked at the Imperial Academy, he had assembled into a single large manuscript the results of all his experiments, including his treasured fluid-flow equation.

Before having it published, however, he wanted to add a concluding section. Just before departing, therefore, he decided to entrust that portion which was already complete to his dearest friend and colleague, Leonhard Euler.

In one last gesture of affection, furthermore, Bernoulli recommended that Euler be appointed his successor as professor of mathematics. Empress Catherine I honored his request, but insisted on appointing Bernoulli, whom she was loath to see leave, a lifetime absentee member of the Academy.

On his return, Bernoulli raced through one country after another, eager as a child for the trip to be over. He was a long way from home and a long way from being a child, however, as he was so pleasingly reminded during the final leg of his journey, on a road just outside of Paris.

He was aboard a horse-drawn coach, conversing with his fellow passengers, when one of them, a botanist, asked for his name. "I am Daniel Bernoulli," the young man replied. Thinking he was being mocked, the stranger said with a sarcastic snort, "Yes, and I am Isaac Newton."

Though Bernoulli repeated his claim, the botanist insisted that his interlocutor was far too young to be *the* famous Daniel Bernoulli. But when Bernoulli showed proof of who he was, the flustered passenger fell silent, remaining star-struck for the rest of the trip.

Bernoulli chuckled to himself. If he was famous now, he thought, wait until the scholarly world had a chance to read his manuscript. It needed only one more chapter to be complete, and then he would have it published.

When finally he arrived in Basel, Bernoulli was greeted like a hero by members of the university's academic senate, by old friends, and by the townspeople. Even from his elderly father, he received a polite greeting and an invitation to stay at his house.

It did not take long for young Bernoulli to readjust to life in his hometown. The climate was kind and so was fate. As professor of anatomy and botany, he had to give lectures, which he loved to do, and most importantly, he had plenty of time to work on his manuscript.

In 1734, just when it seemed that things were going swimmingly, however, Bernoulli's paradisiacal homecoming turned nightmarish. That was the year he and his father were selected co-winners of the French Academy competition.

Although each had won first prize once before, it pained the father to admit that at such a young age, the son was equaling, and probably even surpassing, him as a mathematician. In return, the son was tactless enough not to conceal his youthful arrogance.

The Academy's glad tidings, therefore, ended up degenerating into a colossal collision of conceits, after which young Bernoulli relocated to an apartment of his own and buried himself in work. During the day, he lectured and held meetings with students and faculty; during the night, he worked on his beloved fluids, completing the manuscript by year's end.

In anticipation of this moment, Bernoulli had arranged for the manuscript to be printed in Strasbourg, France, the city where 300 years earlier Johann Gutenberg had invented movable type and that was now famous for its publishing houses. They used the most sophisticated presses in existence; but even so, printing and binding in those days were very slow processes.

For that reason, it took more than *three years* to complete the job. It wasn't until 1738 that Bernoulli finally was able to hold in his hands the printed and bound labors of his adult life. As he opened the cover, his eyes began to water, for there it was, emblazoned on the frontispiece: *Hydrodynamics,* by Daniel Bernoulli, Son of Johann.

He had chosen once again to identify himself in that humble fashion to avoid igniting another confrontation with his father, to prove that Daniel Bernoulli was not the ungrateful son his father had accused him of being. It was meant to be a loving tribute to his father's legacy and fame, but it was to end up being a tribute to the Bernoullis' legacy of bickering and backstabbing.

The tragic turn of events began the very next day, when an excited young Bernoulli dispatched several copies of his new book to his trusted friend Euler. He instructed him to keep one copy for himself and distribute the rest to various important colleagues there in St. Petersburg, including the new empress, Anna Leopoldovna. "Please ask Her to accept this work of mine as a sign of my gratitude," he wrote obsequiously, "and with assurances that I am most certainly not looking to derive any tangible benefits from this gift."

In fact, he was most certainly expecting to derive benefits from the Academy, even if they were not tangible. In recent years, the fledgling Academy in St. Petersburg had become as prestigious as the venerated old academies in Paris, Berlin, and London. Therefore, his fame could be expected to increase substantially once his book came to the attention of its distinguished members.

After nearly ten months of not hearing anything, though, Bernoulli wrote anxiously to Euler, whereupon his dear friend responded with the worst news imaginable: There had been no reaction to his new book, because the copies of it had not yet arrived!

Astonished, Daniel was so beside himself with anxiety, he pestered Euler unceasingly for the next full year, but to no avail. Finally, in 1740, word came that, at long last, the books had arrived; Bernoulli was dismayed, however, by the faint praise contained in Euler's critique and by the dubious-sounding explanation Euler proffered for the long delay.

More than a year after Bernoulli's book had been printed, Euler explained, Bernoulli's father himself had sent Euler part of a manuscript, allegedly containing original research on moving fluids. The old man had proposed to call it *Hydraulics*.

The existence of this partial manuscript had come as a complete surprise, Euler explained in his letter to young Bernoulli, because his former mentor had never once before mentioned working on such a project. Taking the estimable professor at his word, however, Euler had waited eagerly for the second half of the manuscript.

It finally had arrived in late 1740, whereupon Euler had read it, along with Daniel's book, which by then had arrived from Basel. He had written reviews of both works, Euler concluded, trying not to be swayed by his feelings of loyalty to either man; he hoped he had not hurt young Bernoulli's feelings.

Daniel Bernoulli was floored by Euler's letter. Then three years later, in 1743, he was absolutely devastated when his father's book appeared in print. The elder Bernoulli had instructed the publisher to print the year "1732" on the title page, making it appear that his *Hydraulics* had been written *before* Daniel's *Hydrodynamics*.

Moreover, in the preface was an adoring excerpt from the panegyric review that had been written by Johann Bernoulli's pet pupil, Leonhard Euler: "I was thoroughly astounded by the very fluent application of Your principles to the solution of the most intricate Problems, because of which . . . Your very distinguished Name will forever be revered among future generations."

That was hurtful enough, but when Bernoulli read a little further into Euler's gushing review, his anguish became indescribable: "But You also so distinctly and plainly explained the most obscure and most abstruse question about the pressure which the sides of vessels experience as a result of water flowing through them that there remains nothing more to be desired concerning this rather troublesome matter."

Daniel Bernoulli could never prove it, but he would always suspect his father of plagiarism and his alleged friend Euler of duplicity. "Of my entire *Hydrodynamics*, of which indeed I in truth need not credit one iota to my father," Bernoulli lamented, "I am robbed all of a sudden, and therefore in one hour I lose the fruits of a work of ten years."

Euler had purposely delayed a reply to his repeated entreaties,

Bernoulli was convinced, in order to give the elder Bernoulli the benefit of some extra time to complete his spiteful deed. No doubt, the subterfuge had been Euler's loyal-minded way of repaying the old man for having tutored and touted him all those many years ago, and Johann Bernoulli's way of repaying his successful son for having upstaged and humbled him one too many times before.

"What my father does not claim completely for himself he condemns," a bitter Bernoulli complained to Euler, "and finally, at the height of my misfortune, he inserts the letter of your Excellence in which you, too, diminish my inventions in a field of which I am fully the first, even the only, author."

Daniel Bernoulli would never forgive his father for having robbed him of the glory that came with being the first to discover the fluid-flow equation. Above all, he would never forgive his father for having demolished his boyhood dream of becoming the Isaac Newton of his day.

Following these tragic events, young Bernoulli became angry with God, whose plan for him had turned out to be so mean-spirited. He was disappointed, too, with science, whose incompetence at forecasting the future, *his* future, was now all-too-painfully obvious.

In the end, therefore, despondent over his startling fate and his seemingly pointless career, Daniel Bernoulli decided to take control of his own destiny; he decided to quit mathematics: "I would rather have learned the shoemaker's trade than mathematics. Also, I have no longer been able to persuade myself since then to work out anything mathematical. My entire remaining pleasure is to work some projects on the blackboard now and then for future oblivion."

EPILOGUE

In mythology, humans have always found it easy to fly like birds. In a fifth-century Norse legend, for example, a weapons maker named Wayland fashioned himself a suit of feathers and was able to fly merely by obeying these two simple rules of the air: *"Against the wind shalt thou rise easily. Then, when thou wouldst descend, fly with the wind."*

In reality, however, our earliest efforts to fly like birds always ended in disaster. Throughout the Middle Ages, it was a popular sport for people with homemade wings attached to their arms to jump off high towers. If they were lucky, they escaped with a body full of broken bones.

With the publication in 1680 of Giovanni Borelli's unprecedented mathematical study of human muscle power, the world got its first lesson in how very poorly designed the human body was for flying. "It is clear that the motive power of the pectoral muscles in men," Borelli proclaimed, "is much less than is necessary for flight."

According to Borelli's calculations, humans would need pectorals *twenty* times stronger than normal to be able to lift themselves off the ground, using wings of reasonable size. People's only hope, Borelli concluded, was to lighten their bodies in a manner that enabled them to *float* into the air "the same way as a strip of lead can float on water if a certain amount of cork be attached to it."

Borelli's vision of bodies bobbing around in the air came true in 1783, when the Montgolfier brothers, Etienne and Joseph, became the first to get a hot-air balloon to fly. They did not get very high in their ornately designed paper-and-linen balloon, but they did attract worldwide attention—not to mention scare away the birds!

The theory supporting these balloons was simple enough, namely Archimedes's Law of Buoyancy. The problem was in knowing how to control them. In 1785, two Frenchmen, Pilâtre de Rozier and Pierre-Ange Romain, crashed while trying to cross the English Channel aboard a huge, unwieldy balloon with a mind of its own.

Scientifically speaking, balloons and blimps were called *aerostatic* (the airy analog of hydrostatic), because their weight was supported entirely by the buoyancy of still air. By contrast, vehicles supported by the *movement* of air were called *aerodynamic* (the airy analog of hydrodynamic).

Throughout the eighteenth century, these technical distinctions proved secondary to the human disasters. While some daredevils vainly wrestled the controls of their aerostatic monstrosities, others had even less luck getting off the ground in their aerodynamic contraptions.

In 1742, for example, the Marquis de Bacqueville attached four wings made of starched linen to his hands and feet. As he leapt off the Left Bank of the river Seine, he dropped like a rock and broke his leg when he landed atop a washerwoman's barge.

As the decades flew by and the death count of these would-be aviators soared, human optimism took a nosedive. By the nineteenth century, many wondered whether history was trying to tell us something; namely, that we were destined to live out our existence bound to the earth and never know what it feels like to swoop through the air like an eagle.

"Heavier-than-air flying machines are impossible," declared William Thomson, one of Britain's most famous physicists. Even Thomas Edison, the very embodiment of vision and persistence, was doubtful we would ever fly. "The possibilities of the aeroplane," he concluded pessimistically, "have been exhausted."

If history had taught nineteenth-century citizens to be skeptical about scientific efforts to leave the ground, it also had taught them to be skeptical about scientific efforts to predict the future. Two hundred years earlier, Leibniz's hopeful scheme of using the calculus to divine the exigencies of life had failed so utterly, the French playwright Voltaire had derided it in *Candide,* a wickedly satirical comedy in which Leibniz had been identified with the simpleton Dr. Pangloss.

That was not to say that, by the nineteenth century, *everyone* had given up entirely on Leibniz's facile-minded dream. Indeed, hopes were raised anew when Austrian monk Gregor Johann Mendel discovered the laws of heredity and Austrian psychiatrist Sigmund Freud articulated the tenets of psychoanalysis.

Perhaps, philosophers speculated once again, people's behavior was not too irrational to be prophesied by the rational laws of mathematics and science. By knowing "all the forces that give motion to nature and the respective conditions of all natural beings," enthused a brilliant French mathematician named Pierre Simon de Laplace: "Nothing could be uncertain to such an intelligence, and future and past alike would be open to its vision."

In aviation, too, hopes were raised anew when George Cayley, a young British baronet who as a boy had marveled at the exploits of the Montgolfier brothers, designed a flying machine that did not rely on flapping its wings for lift.

Cayley's aeroplane, as he called it, had a fuselage whose aerodynamic shape was patterned after the hydrodynamic shape of a trout. Attached to its top, like one large immovable wing, was a kite. It wasn't very pretty, but it was the forerunner of today's modern airplane.

At first, Cayley built and tested only *unmanned* gliders. They worked so well, however, that in 1849, he dared putting a young boy inside one of them. Much to Cayley's delight, "it lifted off the ground for several yards."

In 1853, emboldened by his success, the baronet coaxed his coachman into the cockpit of his newest glider and pushed him off a hillside. The flight across the small valley ended successfully, but the pilot had been so traumatized by the experience, he quit on the spot: "I was hired to drive," he shouted hysterically, "not to fly."

Following Cayley's breathtaking success, inventors soon started to add gasoline-powered engines to the fixed-wing flying machines. In essence, the strange-looking vehicles began resembling kites propelled by windmill blades.

In the decades ahead, these motorized airplanes flew—or, more accurately, bunny-hopped—their way past many exciting milestones. The grand finale, however, was to be the Wright brothers' historic flight near the seacoast village of Kitty Hawk, North Carolina.

Wilbur and Orville Wright were owners of a bicycle shop in Dayton, Ohio. They fixed and built bicycles for a living, but for years, they put all their mechanical know-how into building a gasoline-powered airplane.

After completing their new flying machine, they chose to try it out at Kitty Hawk, because of the location's strong and steady sea breezes. Like the mythical Wayland, they believed firmly in the idea that *"Against* the wind shalt thou rise easily."

On December 17, 1903, at 10:35 A.M., his shirt rippling in the wind, an anxious Orville approached the machine, slipped into the driver's sling, and gave his brother the signal to start the engine. In a moment, he was carried across the sand and lifted into the air.

As he flew along the coastline, he grappled with the controls, managing to keep the rickety airplane aloft for twelve seconds. Just that quickly, Orville had piloted himself and his brother right into the history books.

It was the first time a self-powered airplane, controlled by a human, had flown for any significant amount of time. The trip had lasted only a dozen seconds, but as brother Wilbur reflected, having witnessed it from the ground, "the age of flight had come at last."

After all these centuries, history had proven the naysayers wrong. It had also proven a larger truth: Fate very often surprised us, but we in turn were capable of surprising Fate.

Because we still did not understand how an airplane was able to fly, however, we were still far from conquering the skies. The Wright brothers notwithstanding, we were a lot like those early hominids who had used fires ignited by lightning without knowing themselves how to create them.

Beginning back in 1871, however, scientists had begun to build wind tunnels for studying the aerodynamics of wings. Being little more than wide-diameter pipes of fast-flowing air, wind tunnels were reminiscent of Bernoulli's pipes of fast-flowing water.

Typically, engineers would put miniature airplanes inside the breezy tunnels and then toss in some metallic dust, so as to make visible the air currents; in that respect, they were mimicking Leonardo da Vinci, who had used grass seeds to study river currents.

One such engineer was a Russian named Nikolai Zhukovsky. As a six-year-old boy, he had fallen in love with kites, after hearing the story of how, in A.D. 906, the Russians had used kites shaped like giant horsemen to frighten the Greeks into surrendering Constantinople.

As a youngster, he had planned to become a military engineer, just like his father. As fate would have it, however, Nikolai Zhukovsky would end up following in the footsteps of someone completely unrelated and unknown to him: Daniel Bernoulli.

Like Bernoulli, young Zhukovsky loved mathematics and the study of solid objects moving through fluids—kites struggling against the air being one of his favorite case studies. Coincidentally, too, in 1868, Zhukovsky had enrolled in a school near the famous St. Petersburg Academy, only to leave a short while later, because "the lectures are none too good" and the harsh climate was worse.

In the course of his subsequent studies at the University of Moscow, Zhukovsky had learned about the multitudinous achievements and multifarious exploits of the famous Bernoulli family. Their tempestuous story, he had discovered, was as captivating as the subject of kites.

Despite Daniel Bernoulli's disheartening experience with his father, Zhukovsky had read with fascination, he had gone on to win *eight* more prizes from the French Academy, bringing his total winnings to ten! That had remained an all-time record, bested only by Euler, who had won twelve prizes, one less than the thirteen children his wife had borne him.

As for Johann Bernoulli, though he had not won any more prizes, he had continued to disparage his son and posture for posterity. In 1748, nearly blind, asthmatic, and gout-plagued, he had died, believing that somehow fate had cheated him out of his just rewards.

Daniel Bernoulli had died peacefully in his sleep at the age of eighty-two. Euler, who had gone blind by then, had remained so industrious, it had ended up taking one hundred pages just to list the titles of his published works.

While reading all this, Zhukovsky had developed a superficial kinship with Daniel Bernoulli, the result of their both having studied in St. Petersburg and dedicated their lives to the study of fluids. It also had brought home to Zhukovsky just how much had changed in a hundred years.

In his day, Bernoulli had faced the question of how to measure blood pressure. Now Zhukovsky faced a much different question, one that had been raised by the astonishing success of Cayley's gliders: How was it possible for an airplane to fly? What exactly lifted it into the air and, in apparent defiance of gravity, kept it there?

After finishing his schooling, Zhukovsky had been appointed a professor at the University of Moscow, whereupon he had applied himself to answering those all-important questions. Back in 1891, after years of pleading, Zhukovsky even had managed to persuade the university to build him a small wind tunnel.

Now, two years after the Wright brothers' stunning achievement, forty-four-year-old Zhukovsky himself was about to fly straight into the history books. Airplanes were able to fly, he announced in 1905, because of Bernoulli's fluid-flow equation.

In order to understand what Zhukovsky had discovered, one needed only to picture a disembodied, miniature airplane wing inside a wind tunnel whose floor and ceiling were flat. The wing's shape was typical, with a flat bottom surface and a rounded top surface.

The cross section of a typical wing, in fact, looked like the upper half of a reclining teardrop that had been divided lengthwise. Inside the wind tunnel, as in actual flight, the blunt end of the semiteardrop knifed through the wind, while the tapered end trailed behind.

Inside the wind tunnel, the wing inevitably split the onrushing air into an upper stream and a lower stream. The upper stream flowed between the wing's upper surface and tunnel's flat ceiling. The lower stream was bounded by the wing's lower surface and tunnel's flat floor. (In effect, the tunnel's "ceiling" played the role of the top of the atmosphere and the "floor," the actual ground.)

Zhukovsky had noted that the upper air stream was slightly *narrower* than the lower one. That was simply because the wing's upper surface was rounded, ever so slightly pinching off the space between it and the tunnel's ceiling.

According to Leonardo da Vinci's Law of Continuity, Zhukovsky had reasoned, the upper (narrower) air stream whizzed by more quickly than the lower (wider) air stream. It was exactly the same reason a river's waters suddenly sped up when squeezing through a bottleneck.

According to Bernoulli's fluid-flow equation, Zhukovsky had concluded, the lower (slower) air stream exerted *more* pressure than the upper (faster) air stream. That is, the air pressure pushing up on the wing was *greater* than the air pressure pushing down.

The net result? Airplanes flew because the pressure beneath their wings overcame the pressure above them. Put another way, airplanes lifted off the ground because their wings were *pushed up* by the relatively high pressure of the air rushing past their lower surfaces. (Or, equivalently, airplanes flew because their wings were *sucked up* by the relatively low pressure of the air rushing past their upper surfaces.)

In the years ahead, twentieth-century historians would look back at Zhukovsky's extraordinary explanation as the dramatic conclusion of one era and the beginning of another. By understanding, at long last, how airplanes were able to defy gravity, modern aeronautical engineers were able to design flying machines not just with their hands but with their minds.

It had taken our species *millions* of years to build an airplane that flew like a bird, millions of years for us to go from lumbering around caves to hovering above them. Amazingly, however, once we had learned exactly how airplanes flew, it took us only *fifty* years to go from soaring above Kitty Hawk to soaring into space.

Ultimately, the credit belonged to Daniel Bernoulli, whose seminal work in hydrodynamics enabled Zhukovsky and others to get the human species off the ground. Ironically, though, most scientific textbooks quickly fell into the habit of referring to the famous fluid-flow equation simply, but ambiguously, as Bernoulli's Principle.

It was not as if anyone seriously doubted it had been Daniel Bernoulli specifically who had first discovered the equation. Rather, it was as if father and son were destined to keep slugging it out, with the outcome forever doomed to remain up in the air.

$$\nabla \times E = -\partial B / \partial t$$

Class Act

Michael Faraday and the Law of Electromagnetic Induction

*I know of no more encouraging fact
than the unquestionable ability of Man
to elevate his life by a conscious
endeavor.*
—HENRY DAVID THOREAU

This evening, as nineteen-year-old Michael Faraday and his friends walked out of Professor Tatum's house, he paused to marvel at the recently installed gas lamps that now lined Dorsett Street. How quickly his world was changing, he thought, and for the better: Gas lamps had made walking the streets of London at night far safer—indeed, the crime rate had plummeted ever since the city had begun installing the bright new lighting three years ago.

A technological revolution was taking Europe by storm, and Faraday wanted so impatiently to participate in it; that was why he was attending Tatum's lectures. He and the others in the group could not afford to go to a university; they came from poor families but were ablaze with a desire to exceed the subjugating expectations of their highly class-conscious society.

Faraday himself was an apprentice booksmith. Were he to surrender to his presupposed fate, he surely would grow up to do nothing more than make books for members of England's upper crust to read. Thanks to his having a master who empathized with his ambition to overcome his lowly lot in life, however, young Faraday was permitted these occasional evenings to educate himself in the mysteries of the natural world.

Tatum's lecture this evening had focused on Luigi Galvani's bizarre discovery of "animal electricity." Nineteen years ago, in 1791, the Italian anatomist had been experimenting with electrical sparks when he noticed they invariably caused dead frogs nearby to twitch. Most surely, an excited Galvani had concluded, this meant that electricity was the source of all animate existence.

Hearing about this remarkable discovery had been especially interesting for Faraday, because his father had passed away only last week. Now, as the young man prepared to walk home, he wondered whether Galvani had indeed found a way to create life.

Putting on his coat, Faraday exchanged good-byes with his friends and watched as they disappeared into the dimly lit city. It was fall, so the fog was particularly heavy, causing Faraday to hesitate for a moment: His family had just moved to this part of London, and he still needed to reassure himself that he was heading in the right direction.

One certainly did not want to get lost in this neighborhood, he thought with a wan smile. Not to say he was complaining: This was all his father had been able to afford, and now things could get even worse, with only his apprentice's modest stipend to support his mother and younger siblings.

Partway through his journey home, the teenager was approached by a bobby who questioned him suspiciously before allowing him to continue along. Someday, young Faraday muttered to himself, it would be different; he would be a respected natural philosopher and treated like a gentleman by everyone.

After some time, as he rounded the corner onto Weymouth Street, Faraday felt relieved to see candlelight filling the windows of unit 18—it reminded him of the warmth with which he had been brought up. He also felt terribly lonely and sad, however, because it reminded him of how very much he missed his father.

Late that night, as he lay in bed, the young man sobbed, burying his face into the pillow, so the others would not hear. He had loved his father even more than he had come to love chemistry, and that was saying a lot.

Young Faraday remembered the time, before the family had moved to the city, when he had been playing in the loft of their old barn. He had fallen through a hole in the floorboards but had been saved from possible death by the cradling arms of his burly father who, as usual, had been toiling over his anvil down below.

Overcome with grief now, young Faraday resolved to emancipate himself from the servile existence into which he had been born. True, in order to support the family, he would have to continue working at the bookbindery, but in the long run, he was determined to develop his mind and become a chemist.

As he mourned his father's absence and contemplated his uncertain future, Faraday began to feel drowsy. His eyes grew heavy, and his final thoughts turned to Tatum's lecture.

Galvani had believed that electricity was the spark of life. Could he be right? young Faraday wondered. Surely scientists knew so very little about the phenomenon of static electricity that one could not dismiss the idea out of hand.

In his growing sleepiness, Faraday's scientific imagination overtook him. What if last week he had been more scrutinizing when his father had drawn his final breath? There in the darkness, would he have been able to see jagged and luminous fingers of static electricity exiting from his father's dying body? These were morbid questions, he realized, but they excited his budding scientific curiosity.

That night, though the young man eventually fell asleep, something within him had been awakened, had been animated by Galvani's spark. It went beyond questions about his father's passing, beyond the Italian anatomist's ghoulish theory; it was the inklings of a new science.

Before it was fully articulated and accepted, however, the blue-collared booksmith would need to do battle with the ignorance and arrogance of the blue-blooded scientific establishment of his day. It was to be a difficult and dramatic struggle, but in the end, this young son of a blacksmith would electrify the world with his first-class mind and a most shocking equation.

VENI

It was 1791, and the civilized world was in the throes of class struggles the scope of which had never been seen before: Suddenly, in both the Old and New worlds, common people were venturing to improve their status by revolting against the status quo.

In the New World, American colonists recently had drafted an unprecedented "Declaration of Independence" and won their liberty from Britain. Now, in the Old World, the lower-class citizens of France having stormed the Bastille prison in Paris, Louis XVI was acquiescing grudgingly to their complaints by signing a "Declaration of the Rights of Man and Citizen."

Coincidentally, furthermore, working-class people in America and Europe themselves were having to acquiesce to the harsh demands of yet another unprecedented uprising, the Industrial Revolution. In England, for example, textile workers by the thousands already had surrendered to a revolutionary army of steam-powered soldiers.

The Industrial Revolution had started fifty-eight years ago, in

1733, when John Kay had invented the flying shuttle—a device that had so sped up the weaving process that the *spinners* hadn't been able to keep up with the new demand for thread. Then in 1765, James Hargreaves had invented a machine that could spin eight strands of cotton at once; after that, it was the *weavers* who had been unable to keep up.

Next, in 1787, the Reverend Edmund Cartwright had invented the power loom—enabling the weavers to keep pace with the spinners, but creating pressure on the cotton *growers* to produce more raw material. In a few more years, that problem, too, would be solved: Eli Whitney would invent the cotton engine—or cotton 'gin for short—which would remove seeds from raw cotton two hundred times faster than any human being!

By 1791, the Industrial Revolution's high-speed robots had increased productivity and profits to an all-time high. They had done so, however, at the expense of the working classes, who now found themselves being exploited or dismissed by employers who were using the newfangled machinery as a way to get rich quickly.

Even for James and Margaret Faraday, who lived in the countryside, far from England's growing industrial centers, this *coup de main* was hitting close to home. Ever since childhood, James had labored long and hard to become a consummate blacksmith; now, though, his superbly wrought handiworks were being steadily devalued by the increasing availability of machine-made products.

In an attempt to find more business, James moved his family to the village of Newington, nearer to London; he needed desperately to earn more money. In the past, his wife had been able to supplement his income by working part time as a maid, but that was not possible right now, because Margaret was pregnant with their third child.

On September 22, as the leaves on the trees began to fall, she gave birth to a son whom they named Michael. The infant's eyes had not yet seen much, but already his tiny red face screamed and scowled at the social upheavals around him. Though joyous at the new arrival, the Faradays were exceedingly anxious: What would become of this child—to them *all*—if James was not able to find steady work soon?

Their only source of comfort in those desperate days was a passionate belief that Jesus Christ would see them through this crisis, as He had many times before. The Faradays were devoted members of what their son would later describe as "a very small and despised sect of Christians known, if known at all, as Sandemanians."

The founder of their church, the late Robert Sandeman, had been a man who had eschewed fancy religious arguments based on scholarly exegeses of the Bible. "That God exists," he had insisted, "is evident from the intricate contrivances of Nature. Let him who doubts cast up his eyes at the heavens and all doubt must vanish." To him, it had been as plain and simple as that.

Most original Sandemanians were people who had split from the Presbyterian Church of Scotland and the Church of England. The sermons from those churches having become too intellectualized for them, the defectors had created a sect that emphasized the childlike faith that Jesus had demanded of his disciples.

This meant, among other things, that the Faradays were not strong believers in formal education. In 1796, therefore, when they moved to northern London, still in search of a steady income, the children were not pressured even slightly to do well in school; worse still, the school itself, being located in a rundown neighborhood, was not very demanding.

In the years ahead, the only time Michael Faraday's parents took an active interest in his schooling was the day he was going to be punished for constantly referring to his older brother Robert as "Wabert." The teacher sent Robert to buy a whipping stick, but the boy hurried home to tell his mother instead.

Sandemanians believed in corporal punishment, in keeping with the admonition in Proverbs 13:24 that "He who spares the rod hates his son, but he who loves him is careful to discipline him." But punishment was unacceptable at the hands of someone from outside their sect, whom Sandemanians regarded as impure. Consequently, upon hearing Robert's story, Margaret Faraday immediately had her children transferred to another school.

Though he had been spared a beating, the quality and quantity of young Faraday's education went from bad to worse. Not only was the new school inferior to the first, the child himself continued to lack any encouragement from his parents, who were too preoccupied with providing for his physical and spiritual well-being.

"My education," Faraday would lament later, "was of the most ordinary description, consisting of little more than the rudiments of reading, writing, and arithmetic at a common day school." That explained why, for years after his brush with punishment, he continued to mispronounce his older brother's name: "Wabert," he would say, not to be a rascal, but because he didn't know any better.

"My hours out of school," Faraday would recall, "were spent in the streets." On a typical day, he and his rowdy band of friends would roam all over the neighborhood and then settle in for a game of marbles in the alleyway next to his family's ramshackle tenement.

During these years, the Faradays lived on nothing more than several loaves of bread a week—a dole from the English government. Oddly, though, even as their situation worsened, the Faradays remained a happy family.

Indeed, all Sandemanians were never happier than when they were penniless. Poverty reminded them of when Jesus, who, impecunious himself, had warned the Israelites that a rich man had less of a chance to enter the Kingdom of God than a camel had of passing through the eye of a needle.

For that reason, Sandemanians were rugged and unpretentious

people, able to survive on very little, save for their bountiful faith in the mercy of God's only Son. Indeed, since they believed that God had specifically chosen Sandemanians to enter heaven when they died, very little frightened them—apart from the threat of excommunication.

That danger was taken very seriously, which was why young Faraday and the other Sandemanian children were allowed to run wild during the week, but on Sunday afternoons, they were required to be in church. Failing to do so even once, without a proper excuse, was grounds for being expelled from the church.

Despite his being brought up to be a good Sandemanian, in the eyes of contemporary English society, Michael Faraday was little more than a poor, ignorant street urchin. Furthermore, when he was thirteen years old, though he could barely read or write, time had come for him to quit school altogether.

According to the traditions of the working class, the young man now needed to find a job. The routine was all very clearly spelled out: He would begin with some kind of apprenticeship, during which he would acquire a skill that would earn a living for him and whomever he would choose to marry.

Under normal circumstances, James Faraday would have wished for his son to become a blacksmith. But these were anything but normal times, made even more precarious by the fact that England was now at war with France.

During the years young Faraday had been growing up, France's proletariat had guillotined Louis XVI and his queen, Marie Antoinette. Now, in 1804, the bourgeoisie had crowned as their new emperor one Napoleon Bonaparte, an imperialistic general who designed to conquer the world with the aid of new and deadly military machines spawned by the Industrial Revolution.

With promises of providing a nurturing environment for experimenting and inventing, Bonaparte had attracted to France talented young scientists and engineers from all over the world, especially America, whose recent independence could not have been won without French assistance. Indeed, at this very moment, an American inventor named Robert Fulton was tinkering with steam-powered boats on the river Seine.

Clearly, it seemed, the future belonged to steam and to those who exploited its power, for good or ill. Therefore, though it broke his heart to admit it, the elder Faraday knew his son could not—should not—follow in his footsteps.

Fortunately, the news was not all bad. As automation had increased, so had the opportunities for unskilled laborers—poor, uneducated young people just like Michael Faraday. He had many job opportunities, therefore, and he thought about them all very carefully.

Unable to choose, though, young Faraday made an interim decision that would prove to be as consequential as it was ironic: This semiliterate boy from the wrong side of the river Thames decided to become the new errand boy at a nearby bookstore run by a Mr. George Riebau.

The job did not require him to read. In fact, all he really needed to do was scurry around the neighborhood, something at which the former waif was quite experienced. It did not pay much, he had judged, but it was preferable to being cooped up inside one of those depressing and hazardous sweatshops that were now sprouting up all over the city.

As it turned out, the job was desirable for still another reason. Unbeknownst to Faraday, the literacy rate was skyrocketing throughout industrialized Europe, in part because mechanized printing presses and boats were making it easier and cheaper to produce and distribute books. Consequently, people were buying them in record numbers, keeping the new delivery boy very busy.

Intrigued by the widespread interest in books, young Faraday's own attitude toward the printed word slowly began to change. This miraculous transformation was abetted, furthermore, by his becoming interested in what went on at the back of Riebau's shop; there pages of text were bound together into books.

This aspect of the business so attracted young Faraday that in 1805, he made the decision to become an apprentice. For the first time in his fourteen years, he had quit the streets. He had never once gone to a library, but for the next seven years, a library of books from all over the world would come to him.

As a novice, Faraday had a lot to learn, and the work was not easy. Bookbinding had been one of the few crafts to survive the Industrial Revolution, precisely because it required a mental concentration and manual dexterity that no clumsy machine—and very few people—could bring to the task.

Faraday was taught to take pages from a printer, stitch them, trim them, and secure them to a handmade leather cover. The process demanded scientific precision; the product was a work of art.

Though the young apprentice was astonished at the effort that went into making a book, he was equally surprised to discover how difficult it was to read one. Increasingly, he became frustrated and angry by his inability to enjoy the fruits of his labor—like the construction worker who discovered he was unqualified to attend a college he had helped to build.

Consequently, the young teenager began to teach himself how to read. It was a laborious and painful process, but in a matter of months, he made up for what he had neglected to learn during all those years of public schooling.

One day, while stitching together the newest edition of the *Encyclopaedia Britannica,* Faraday's life changed forever. In reading its 127-page entry on electricity, Faraday learned that, though natural philosophers had known about this invisible phenomenon for centuries, they still had not figured it out.

Something within him stirred, reminding him of a biblical verse he had heard a thousand times before; it was from Romans 1:20:

"Since the creation of the world, God's invisible qualities—his eternal power and divine nature—have been clearly seen, being understood from what has been made."

So long as electricity remained invisible and enigmatic—which was to say *not* "clearly seen" and *not* "understood"—it would be impossible for anyone to have a proper understanding of "God's eternal power and divine nature." This was intolerable, the young Sandemanian decided, so then and there, he resolved to help remedy the situation.

Because he had been brought up to believe in the fundamental simplicity of man's relationship to God, Faraday doubted that electricity could be as complicated as all that. Fortunately, the London of his day offered this unsophisticated youngster unparalleled opportunities to find out for himself.

In recent years, the Industrial Revolution had caused such a widespread interest in science and technology that natural philosophers had begun to write popular magazine articles and books and to conduct lectures especially for the public. The books were snatched up as quickly as they were published, and the lectures were usually delivered to standing-room-only crowds.

For Faraday, the unprecedented demand for science books was a double blessing: As a bookbinder, it meant job security; as a would-be natural philosopher, it meant finding plenty of information about electricity written in plain English. "It was in those books, in the hours after work," Faraday would recall later, "that I found the beginning of my philosophy."

While he rejoiced in the books that were freely available to him, however, Faraday was chagrined about being too poor to buy tickets to any of the public lectures, particularly those of Humphry Davy, the famous chemist and director of the prestigious Royal Institution of London. In recent years, Davy's extravagant and lively presentations had received such rave reviews they had become legendary.

The youngster had become possessed by a desire to see Davy in action, a reasonable wish, considering that the Royal Institution was located only a short distance from Riebau's shop. As English society saw it, however, this presumptuous apprentice might as well have been wishing to visit an enchanted castle in some faraway land.

In nineteenth-century England, science was not yet a paying profession, so the only people who could afford to do it were the very wealthy. The Royal Institution was like an ultra-exclusive country club, and its aristocratic members never would have deigned to consort with the likes of Michael Faraday or anyone else of his lowly class.

It didn't stop there: Even more elitist than the Royal Institution was the Royal Society, located nearby; its members were the equivalent of scientific royalty. Therefore, Faraday's wanting to become a scientist was akin to a pauper dreaming about becoming a prince.

Mercifully, the young proletarian was not old enough to understand any of this, and his master Riebau was too kind-hearted to disabuse him of his fantasies. Indeed, Riebau so empathized with the boy's desire for a better life that he gave in to Faraday's pleading request to convert part of the shop into a makeshift laboratory.

After hours, Riebau's fireplace became Faraday's furnace and the mantelpiece his workbench. The young man's lab equipment was crude, of course, but doing the experiments and keeping careful notes in his journal made him feel like a real philosopher.

In the months ahead, the teenager built himself an electrostatic generator, a hand-cranked device that created sparks of static electricity. He also managed to save up a few shillings to buy a pair of Leyden jars; static electricity was by its nature elusive, but it could be trapped and stored inside these jars, like fireflies in a bottle.

Faraday also began reading self-help books, because he realized that if he was to become a man of science, he would have to learn not just its theories but its techniques, as well. In *Improvement of the Mind,* a book by Dr. Isaac Watts, for example, Faraday learned the four best ways of becoming smarter: attend lectures, take careful notes, correspond with people of similar interests, and join a discussion group.

In 1810, unable to afford showy public lectures, Faraday joined a discussion group consisting mainly of working-class young men who aspired to improve their stations in life. Every Wednesday night at eight, with Riebau's permission, Faraday would leave work and travel to the house of a science teacher named John Tatum.

During these meetings, either Tatum or one of the attendees would deliver a lecture on a subject of his choice. Faraday always would listen intently and take careful notes; at the end of it all, he planned, he would bind all his notes together into one big, beautiful book.

When Faraday's turn came to give a presentation, he spoke about electricity and drew a warm and enthusiastic response from his confreres. Tatum wasn't Davy and his house wasn't the Royal Institution, but at only one shilling per week, the meetings were eminently affordable and enlightening.

In the course of learning to become a natural philosopher, Faraday revealed himself to be as distrustful in matters scientific as he was trustful in matters religious. Whereas he accepted literally and unquestioningly anything written in the Holy Bible, he put to the test any assertion made in books written by mere mortals.

"In early life I was a very imaginative person, who could believe in the *Arabian Nights* as easily as in the *Encyclopaedia,*" Faraday recalled later, "but facts were important to me, and saved me. I could trust a fact, and always cross-examined an assertion."

To Faraday, facts were as sacred as scriptural verses, in that both were the only reliable ways of comprehending God's creation. Each evening, therefore, after everyone else had left, the young man would sequester himself in Riebau's shop-*cum*-lab and duplicate every experiment mentioned in the books he had read. "I was never able to make a fact my own," he would confess, "without seeing it."

Faraday had never felt so invigorated as he did now, but the same could not be said about his long-suffering father back home. Recently the elder Faraday had written to Michael's brother Thomas to complain that: "I am sorry to say I have not had the pleasure of enjoying one day's health for a long time."

The doctors were puzzled as to what was debilitating the father, but their prognosis was that he would become an invalid. One last time, therefore, the Faradays moved to another tenement, located more conveniently near the center of town. Within months, however, James Faraday died.

In the two years that followed, Michael Faraday helped support his mother and siblings, all the while nursing the hope of becoming a natural philosopher. But in 1812, the year his apprenticeship was scheduled to end, ruefully, he began surrendering to the likelihood of having to live *down* to the expectations of his society: Unless something miraculous happened to change the course of his life, it appeared he was doomed to become a bookbinder, just like his master Riebau, and to keep science merely as a hobby.

As winter was coming to an end, however, a man named Dance Junr walked through the front door of the bookshop—and into Faraday's life. During his last visit, Junr had spotted the ornately crafted book young Faraday had made of his notes from Tatum's lectures. Curious as to its contents, Junr had asked and received permission from Riebau to borrow it for a while.

Now, several weeks later, he was returning it directly to Faraday, with four small pieces of paper sandwiched between the pages. Junr was a member of the Royal Institution, Faraday was astonished to discover, and out of admiration for the young man's work, he was returning the book, along with complimentary tickets to the next series of public lectures by the famous Humphry Davy!

VIDI

Long before Christians had come to believe in the Father, Son, and Holy Ghost, natural philosophers had stumbled on their own trinity: electricity, magnetism, and the gravitational force. These three forces alone had governed the creation of the universe, they believed, and would shape its future forevermore.

Their belief had been founded on a rock, literally, 600 years before Christianity. Back then, the Ionian philosopher Thales of Miletus had noticed that lodestones attracted shards of iron and that amber—fossilized tree sap—attracted chaff and bits of straw, after being rubbed with wool. Added to those mysterious forces was the self-evident fact that the earth attracted objects of *all* kinds.

Given the forces' disparate behaviors, it was no wonder that philosophers very early on were left scratching their heads: Were these three forces completely different? Or were they, like the Christian Trinity, three aspects of a single phenomenon?

They were tempted to believe in the unity of the three forces, simply because it was most consistent with their notion that, despite its ostensible complexity, Nature fundamentally was simple. Unfortunately for that tidy premise, however, every shred of evidence indicated the three forces were indeed as varied as their outward behavior implied.

Ancient philosophers venerated gravity above the other two forces, because it alone appeared to be universal; it was everywhere, at all times. Ultimately, too, gravity's influence was irresistible: It had the power of felling mighty trees and mighty kings.

In comparison, the lodestone and amber forces did not have anywhere near as conspicuous a presence in people's everyday lives; indeed, lodestones were quarried in only a few places on earth, and amber was as rare as gold. Furthermore, these forces seemed to influence only specific things, and only under very specific conditions. In short, compared to gravity, they were a novelty, a mere curiosity.

Not surprisingly, therefore, in his celebrated magnum opus, *Physics*, even the widely inquisitive Aristotle made no mention of Thales' two eccentric forces. He did, however, call attention to the mystery of "the natural downward motion of the earth" and, in referring more than once to the "gravity" of solid objects (in contrast to the "levity" of gaseous objects), gave the ubiquitous force its name.

During the centuries that followed, the slight persisted: Soberminded philosophers studying gravity did not allow themselves to be distracted by minor mysteries, such as lodestones and amber. In the society of ideas, as it were, gravity had become a pampered aristocrat who lorded over a pair of nameless nobodies.

It wasn't until 1581, when an English physician became interested in their possible curative powers, that Thales' two ill-treated forces began being taken seriously. His name was William Gilbert, and for years, he went around rubbing everything in sight with wool, silk, and fur; it made his colleagues wonder about his sanity, but in the end, the good doctor discovered something truly amazing.

Gilbert had been able to elicit Thales' amber force by rubbing diamonds, sulfur, sealing wax, and many other ordinary substances, not just amber. Moreover, the force attracted not only chaff and straw but "all metals, woods, leaves, stones, earths, even water and oil, and everything which is subject to our senses, or is solid."

Indeed, since it now appeared that the amber force was nearly as universal as gravity, Gilbert decided it deserved a name of its own. He christened it the *electrick* force, after *elektron,* the Greek word for amber.

Gilbert was just as enthusiastic about lodestones. "The loadstone far excels all other bodies known to us in virtues and properties pertaining to the common mother [earth]," he raved, "but those properties have been far too little understood or realized by philosophers."

In this case, Gilbert was *not* the first to name the phenomenon; Thales' lodestone force already had come to be called *magnetism,* after the Asian district of Magnesia, where many lodestones were mined originally. Gilbert *was* the first, however, to discover that the two ends of a magnet always behaved differently; he called them the *north pole* and the *south pole.*

According to Gilbert's experiments, whenever a pair of magnets was brought together, similar poles *repelled* one another and dissimilar poles *attracted.* That is, two neighboring magnets always spun around so that the south pole of one magnet lined up with the north pole of the other.

Gilbert wondered whether this surprising discovery might help explain why a magnetic compass needle always pointed northward. Up until that time, natural philosophers had speculated that compass needles behaved that way simply because they were attracted to the North Star or to some fabled lodestone-laden mountain range located within the Arctic Circle.

After giving it some thought, Gilbert presented "to the view of the learned our New & unheard of doctrine." The entire earth was a magnet, he proposed, complete with its own two poles! That explained a compass's behavior: Its magnetic needle's south pole naturally was drawn to the magnetic earth's north pole (and viceversa, the needle's north pole was drawn to the earth's south pole).

Gilbert published these revolutionary observations and theories concerning Thales' two forces in a book titled *De magnete, magnetisque corporibus, et de magno magnete tellure* (On the magnet, magnetick bodies also, and on the great magnet the earth; a new Physiology, demonstrated by many arguments & experiments). It was a seminal book, because Gilbert was the first to make use of what came to be called the *scientific method,* that unique and powerful blend of speculation and experimentation.

This remarkable publication disappointed Gilbert's colleagues, however, because it seemed to dash any remaining possibility that electricity, magnetism, and gravity were somehow related. Already philosophers had known that gravity was different from the other two forces; now, according to Gilbert's revolutionary observations, those two forces themselves were complete opposites.

Whereas electricity was *sympathetic* (strictly attractive), magnetism was *polar* (attractive and repulsive); and whereas electricity was *catholic* (attracting many different kinds of things), magnetism was highly *selective* (affecting only iron and other magnets). In the early seventeenth century, therefore, things looked gloomy for disciples of the scientific belief that simplicity underlay the natural world.

In 1663, they were heartened to hear of an experiment done by a German named Otto von Guericke. After rubbing a piece of sulfur with his bare hands, he had discovered it attracted many things, just as Gilbert had found, but it also *repelled* certain things!

Evidently, electricity was not so completely different from magnetism after all. If von Guericke was to be believed, both forces were capable of being attractive *and* repulsive, which made philosophers wonder anew about gravity: Was it possible that their pet force, too, was capable of *repelling* objects?

If gravity were able to push things away, then a person would expect to see objects floating up into the sky, entirely on their own. The possibility seemed to defy gravity and common experience, and yet, philosophers enthused, no one had ever proven that it could not happen.

It was not until 1687 that the famous English natural philosopher Isaac Newton brought science back down to earth by publishing his monumental three-volume work *Philosophiae Naturalis Principia Mathematica* (Mathematical Principles of Natural Philosophy). In it, he offered compelling evidence that gravity worked in only one way—it always pulled objects to itself, never pushed them away. (See "Apples and Oranges.")

With that matter apparently settled, science returned its attentions to von Guericke's sulfur experiment and the striking similarity in the behavior of electricity and magnetism it had revealed. Were the two forces similar in other respects? The answer, natural philosophers soon began discovering, was a resounding yes.

In 1785, for example, a Frenchman named Charles-Augustin Coulomb hung small bar magnets from the ends of strings and watched how they influenced each other when separated by various distances. He discovered that the force between them diminished with the *square* of their separation: If the distance between magnets increased twofold, the force decreased fourfold ($2^2 = 4$); if the distance increased threefold, the force decreased ninefold ($3^2 = 9$); and so forth.

This revelation was especially remarkable, because by also suspending *electrically* charged objects on strings, Coulomb discovered the electric force obeyed an identical rule! By the end of the eighteenth century, therefore, science was beginning to feel sanguine about the possibility of finding some modicum of unity between at least two of Nature's three forces.

At the same time, however, science still was not sure how to make use of electricity and magnetism, unified or not. Whereas Newton's theory of gravity was now leading to all kinds of useful predictions—such as the moon's gravitational influence on the tides, the existence of new planets, and the flow of water over miles-long aqueducts—electricity and magnetism were more entertaining than enlightening.

Back in 1745, for example, Dutch physicist Pieter van Musschenbroek had invented a special jar—named after the University of Leyden—that was able to store large amounts of electricity. The invention of the Leyden jar, which foreshadowed the modern battery, meant that natural philosophers no longer had to settle for tiny amounts of static electricity eked out by stroking "elecktrick" materials such as amber.

During an early demonstration of his novel creation, van Musschenbroek had felt its frightening sting. It had happened as he was
reaching for a Leyden jar filled to the brim with static electricity:
"The arm and body was affected in a terrible manner which I cannot express," he later reported, "in a word, I thought it was all up
with me." Van Musschenbroek had discovered the electric spark.

At first, the showy new phenomenon commanded some "ah-
has" from serious-minded philosophers, but soon it began eliciting
many more "oohs and aahs" from public audiences all over the
world. Natural philosophers attempting to satisfy the public's
growing interest in science and technology had found that Leyden
jars in particular—and electricity and magnetism in general—were
real crowd pleasers.

No one knew that better than Luigi Galvani, who back in the
1770s had begun a series of electrifying experiments intended to be
scientifically serious. He and some assistants had gathered around a
freshly dissected frog and what they saw had made them jump out
of their skins: "Now when one of the persons who were present
accidentally and lightly touched the inner neural nerves of the frog
with the point of a scalpel, all the muscles of the leg seemed to
contract repeatedly as if they were affected by powerful cramps."

After having given some thought to what he had just seen, Galvani had leapt to the conclusion that he had found the spark of life,
an occult assertion that ultimately had ostracized him from the
scientific community. Now, many years later, the disreputable
Galvani had turned into something of a carnival barker, and electricity into a seedy sideshow.

During his sensational public lectures, Galvani showed people
how dozens of frogs' legs twitched uncontrollably when hung on
copper hooks from an iron wire, like so much wet laundry strung
out on a clothesline. Orthodox science cringed at his theories, but
the spectacle of that chorus line of flexing frog legs guaranteed
Galvani sell-out crowds in auditoriums the world over.

The provocative Italian managed to make a believer even out of John Wesley, the learned Englishman whose founding of Methodism during the early eighteenth century had inspired fellow evangelist John Sandeman to create his own sect. Wesley had learned about Galvanism—the name given to Galvani's theory of animal electricity—during his student days at Oxford; now, in 1784, he decided to include in the dissident religion's new constitution the explicit conviction that "electricity is the soul of the universe."

In 1800, the electricity sideshow suddenly acquired an even greater power to attract large and distinguished audiences. It was the result of yet another flamboyant Italian physicist, this one having invented a device that could produce a steady *flow* of electricity, a vast improvement over Van Musschenbroek's fleeting sparks and Galvani's momentary twitches.

His name was Alessandro Volta, and his invention consisted of stacking, like poker chips, disks of copper and zinc, separated by disks of cardboard moistened with salt water. They were called Volta's "piles," because thirty to sixty disks had to be piled up in order to produce a measurable effect; the higher the pile, the stronger the electric current.

Volta's piles were, in fact, history's first modernlike storage batteries. To determine if one was working, a person needed only to touch its two terminals to the tip of the tongue; the electric current—too weak to harm anyone—invariably created a tingling, sour sensation. (The very same effect can be produced by touching the tongue with a silver spoon and a piece of tin foil connected to it.)

On March 20, 1800, an enthusiastic Volta wrote to Joseph Banks, president of the Royal Society of London, informing him of a horizontally arrayed version of his pile: "We set up a row of several cups . . . of pure water, or, better, of brine or of lye. We join them all together in a sort of chain by means of metallic arcs [which bridge adjacent cups of liquid]."

Volta called this his "crown of cups," and Banks was so impressed by it that he showed the letter to his colleague William Nicholson, a civil engineer working in London. Nicholson and his aristocratic colleague Sir Anthony Carlisle replicated Volta's new device immediately and within a month made a jolting discovery of their own.

When Nicholson and Carlisle had taken the two wire terminals from their crown of cups and dipped them into a container full of water, the water had begun to bubble. At first they had been puzzled, but then they had concluded the electric current was somehow decomposing the water into its two basic elements, hydrogen and oxygen; both were gases, which explained the bubbling.

No one understood exactly how an electric current was able to produce this effect, but it resembled the behavior of a lightning bolt, cleaving whatever objects it hit. In any case, the mysterious phenomenon was undeniably real and eventually given the name *electrolysis,* from the Greek "to loosen with electricity."

Just that suddenly, science had discovered a reason for taking electricity seriously: The entertaining force had a useful purpose, especially for chemists. Recently they had embraced the novel idea that matter consisted of only a few dozen essential elements; here, now, was the perfect means of putting their idea to the test, for prying those alleged atoms out into the open.

Immediately hundreds of chemists everywhere set upon building themselves Voltaic piles or crowns of cups in hopes of being the first to discover some new atomic element. Among them, one chemist emerged as the most proficient at applying this new technology to his subject: Humphry Davy.

By 1807, five years after having arrived at the Royal Institution in London, Davy had built one of the world's largest and most powerful Voltaic piles and used it to isolate two hitherto unknown elements: sodium and potassium. A year later, he used his pile to discover four more elements: barium, boron, calcium, and magnesium. It was an achievement so remarkable that Napoleon, though he was at war with England, awarded to Davy the *Institut de France*'s prestigious Bonaparte prize.

From then on out, electricity and magnetism were destined to bring new meaning to the emerging science of chemistry. Reciprocally, chemistry was to bring new credibility to the study of electricity and magnetism, and something more—the extraordinary talents and ambitions of a young man named Michael Faraday, who at this very moment was coming of age in London.

Like Thales' two ancient forces, young Faraday always had occupied a gallingly humble place in the scientific community. But now, after all these years, the would-be scientist was about to become the very ticket this would-be science of electricity and magnetism needed in order to become a first-rate discipline.

VICI

On February 29, 1812, Faraday leapt up the stone steps and rushed through the heavy doors of London's Royal Institution. Symbolically, it was like the storming of Bastille, except it was Faraday who would lose his head, not any of the noblemen who dwelt within.

This was the evening for which he had been waiting so long. For years, he had fantasized about this palace of science. And now, as he gawked and walked his way through the opulent antechamber and into the auditorium, Faraday nearly swooned at the reality of it all.

Once seated, the young man opened his notebook and began drawing and describing the elegant room and the gadget-filled stage in front of him: "literary and scientific, practical and theoretical, blue-stockings and women of fashion, old and young, all crowded—eagerly crowded—the lecture-room.''

The event was scheduled to start at eight o'clock, and at that moment precisely, all eyes turned to watch the tall, handsome lecturer stride onto the dais. Humphry Davy was not a king, but he bore himself like one. To many—not just to the young bookbinder applauding adoringly from his seat in the center section—he was one of the greatest natural philosophers of his day.

When the applause quieted down, Davy proceeded to dazzle the gallery with his legendary talents and fantastical demonstrations. Chemicals glowed, electricity flowed, and throughout it all, Faraday himself glowed, and the ink from his pen flowed; by the time it was over, the eager young man had filled up ninety-six pages of notes and illustrations.

For the attendees, it was the end of a memorable evening, made all the more so by rumors that this was to be Davy's final series of lectures. For the twenty-year-old Faraday, however, it was to be the beginning of a revolutionary scientific career, one that ultimately would lead to the dethronement of the Royal Institution's vaunted liege.

As the ecstatic young Faraday walked home, his lighthearted mood gradually was overcome by the surrounding darkness. His apprenticeship would expire in only eight months, he thought glumly, at which point he was committed to work as a journeyman for the French bookbinder Henri de la Roche. The wages would be enough to support him and his widowed mother, but the job itself would not make him happy.

That evening, Faraday had come within arm's length of his dream, the closest he ever had gotten to it; now, more than anything, he wanted to grab hold of it. But how could someone as insignificant as he seize Davy's attention?

During the next several months, while the increasingly anxious young man attended Davy's three remaining lectures, an idea came to him. He would recopy his lecture notes and make them into a book so exquisite, Davy would be sure to notice it—and him. His book of notes from Tatum's lectures had gotten him into the Royal Institution, Faraday reasoned; perhaps this one would get him employed there.

No sooner had he congratulated himself for having come up with such a brilliant game plan, however, than a public announcement was made: Within these past few days, Humphry Davy had been knighted by the queen and betrothed to a wealthy widow. The couple was now honeymooning in Scotland until the end of the year.

Faraday was beside himself with anger and frustration: He could not wait that long, because by the end of the year, his fate as a bookbinder would be sealed! In desperation, Faraday wrote to Sir Joseph Banks, president of the Royal Society; the young man implored him to help but received not so much as a reply.

On October 7, Faraday's apprenticeship came to an end, along with any hopes for a better future. The following day, he reported to his new job and immediately took a disliking to his boss. Monsieur de la Roche was quick-tempered and, worse, made it clear he was *not* going to indulge Faraday's scientific pipe dreams, as Riebau had done for so many years.

As fall turned to winter, Faraday's precious memory of Davy's spring lecture series began to shrivel and die like the leaves all around him. "I am working at my old trade, the which I wish to leave at the first convenient opportunity," he wrote despondently to a friend, "Indeed, [unless] I stop in my present situation I must resign philosophy entirely to those who are more fortunate in the possession of time and means."

In December, having found out that Sir Humphry and his bride had returned to London, a critically unhappy Faraday decided to follow through with his original plan: "My desire to escape from trade . . . and to enter into the service of Science . . . induced me at last to take the bold and simple step of writing to Sir H. Davy," he later would recall, "at the same time, I sent the notes I had taken of his lectures."

In the days that followed, the young man awaited a reply, but none came. Then, on December 24, an elegantly dressed footman appeared at 18 Weymouth Street. He knocked on the door of the Faradays' dilapidated apartment and handed to Michael this note from the Royal Institution's monarch himself:

> I am far from displeased with the proof you have given me of your confidence, which displays great zeal, power of memory, and attention. I am obliged to go out of town, and shall not be settled in town till the end of January. I will then see you at anytime you wish. It would gratify me to be of any service to you; I wish it may be in my power.

Faraday felt as giddy as all those children throughout London who were eagerly awaiting the imminent arrival of Old Father Christmas. He had waited a lifetime for this opportunity, and now he needed to wait only a month longer; that month, however, seemed to last an eternity.

When the big day finally did come, Faraday's meeting with Davy passed so quickly he wondered if it had been only a dream. He remembered having felt faint while shaking Davy's hand, then hopeful as the scientific nobleman had listened to his pleas for employment, and ultimately devastated, when Davy had explained he had no jobs to offer and that Faraday would be wise to keep his present position as a journeyman bookbinder.

As he tottered down the steps of the Royal Institution, the young man was certain he would never pass through its doors again. All that effort, all those big plans, all of that anticipation: It all had come to naught.

For months, there had been bad blood between Davy's assistant and another employee at the institution. The two had managed to maintain their civility, but several weeks after Faraday's visit, their simmering feud suddenly exploded into a flurry of blows.

On the morning of March 1, as Faraday prepared for work, there was a knock at the door. It was the footman again, with the message that Davy's assistant had been fired for fighting.

If he was still interested, Davy offered, Faraday could have the job *and* a small, two-room apartment above the lab. Still interested? Without waiting to reread the message, Faraday began packing, and shortly afterward he rushed out the door to inform his boss.

To Faraday's surprise, Henri de la Roche had come to like him. "I have no child," the hot-tempered Frenchman now confessed, "and if you stay with me you shall have all I have when I am gone." Faraday, however, was as fanatical about becoming a natural philosopher as he was about being a good Sandemanian—nothing, and no one, could change his mind.

Within minutes, Faraday bounded into the Royal Institution, hardly believing that this was now to be his home as well as his place of employment. He felt like a frog-turned-prince and was unfazed as Davy explained to him that being a lowly lab assistant involved merely washing test tubes and sweeping the floor.

"He still advised me not to give up the prospects I had before me, telling me that Science was a harsh mistress . . . poorly rewarding those who devoted themselves to her service," Faraday would recall later. "He smiled at my notion of the superior moral feelings of philosophic men, and said he would leave me to the experience of a few years to set me right on the matter."

On the contrary, during the next few years, the young lab assistant reveled in the service of science. Among other things, he learned to extract sugar from beetroot, to improve the chemical properties of steel, and to use electrolysis to disassemble a variety of compounds.

It was as if he had become an apprentice all over again, except this time the object of his handiwork was the great book of Nature:

how it was stitched together, and how it might be understood by science and improved by technology.

In the process, Faraday learned how to survive the physical dangers that came with working in a chemistry lab. "I have escaped (not quite unhurt) from four different and strong explosions," he reported to a friend.

> Of these the most terrible was when I was holding between my thumb and finger a small tube containing 7½ grains of [nitrogen trichloride]. The explosion was so rapid as to blow my hand open, tear off a part of one nail, and has made my fingers so sore that I cannot yet use them easily.

During his first trip abroad—which began in October 1814—Faraday also learned how to survive the insults that came with being a working-class bookbinder who sought acceptance into the high-class world of science. In some ways, the barbs were even more difficult to endure than the chemical explosions.

The foreign scientists were not the problem: They all fell in love with this unassuming young man who was so easily intoxicated by anything and everything scientific. The culprit was Davy's wife . . . and to some extent, Davy himself.

Originally, Davy had invited Faraday to join him on the research and lecture tour as his *lab assistant*. However, because the ongoing Napoleonic wars had made travel in Europe very dangerous, Davy's fretful valet had backed out of the trip at the very last minute.

Reluctantly, Faraday had agreed to double as Davy's manservant—but only as far as Paris, their first stop, whereupon the aristocratic chemist had promised he would find someone else for the job. In fact, Davy never did find a valet who could satisfy his snooty standards, so for the entire trip, he required Faraday to be his flunky-in-waiting as well as his lab assistant.

That was the injury; Lady Davy was the insult. "She likes to show her authority," Faraday complained in a letter to a friend, and is "extremely earnest in mortifying me." She saw how competently Faraday assisted her husband in research, yet she insisted on introducing the young man to everyone as their *servant* and on treating him in a manner befitting that position.

The humiliating trip, however, was not a complete disaster for Faraday. Owing to his mentor's world-class stature, he was able to meet and work with some of Europe's finest scientists—including Alessandro Volta, who had become quite a celebrity since his invention of the battery, and Andre-Marie Ampère, a middle-aged Parisian prodigy who was wowing the world with his phenomenal mathematical abilities.

These were scientists about whom Faraday had read during all those years as an apprentice in Riebau's shop. These were the scientists whose work he had tried to duplicate in his makeshift lab, using crude and inexpensive apparatus. Now, to his wonderment, he was able to speak to them and to inspect firsthand the elegant and expensive equipment they had been using to investigate electricity, magnetism, and other natural phenomena.

"I have learned just enough to perceive my ignorance, and, ashamed of my defects in everything, I wish to seize the opportunity of remedying them," Faraday wrote in a letter midway through his trip, "the glorious opportunity of improving in the knowledge of chemistry and the sciences continually determines me to finish this voyage with Sir Humphry Davy."

By the time he returned to London, in the spring of 1815, Faraday had accumulated what resembled an upper-class education: After completing public school and a few years at Oxford or Cambridge, young British aristocrats of the day usually toured the Continent, accompanied by their tutors. Therefore, although socially Faraday still belonged to the lower classes, professionally he now was positioned to earn himself a respectable place in the scientific community.

Within days of their return, a grateful and somewhat sheepish

Davy rewarded Faraday with a dual promotion to the positions of Superintendent of the Apparatus as well as Assistant in the Laboratory and Mineral Collection. The elder chemist also encouraged Faraday to start experimenting on his own, which he did, beginning with a rock sample he had collected while in Italy.

In 1816, Faraday published his results—"Analysis of Native Caustic Lime of Tuscany"—in the *Quarterly Journal of Science*. It was his first scientific publication and something of a declaration of independence: Officially now, he was free of being Davy's puny protégé.

In the years ahead, Faraday's emergence as a gifted scientist shook the Royal Institution like a vial of exploding nitrogen trichloride. Now that he had access to proper equipment, he revealed himself to be a technical wizard—some even began to say Davy's heir apparent.

Faraday assembled experiments the way he once had put together books, with extraordinary patience and precision. He had such a keen eye for detail, furthermore, that scientists tended to accept his word for the existence of this-or-that subtle effect, even if they themselves had not yet seen it with their own apparatus.

Faraday was unlikely to return the compliment, however, because he had not lost any of that uncompromising skepticism he had evidenced as a youngster. Time and again, he would refuse to agree on the existence of some phenomenon until he had seen it for himself, explaining: "The philosopher should be a man willing to listen to every suggestion, but determined to judge for himself . . . He should not be a respector of persons, but of things. Truth should be his primary object."

At the same time, his religion and social standing in life had made Faraday a humble man. When expressing his vaunted skepticism to a colleague, therefore, he was careful not to become too smug about his own abilities or ideas: "By adherence to a favourite theory, many errors have at times been introduced into general science which have required much labour for their removal . . . To guard against these requires a large proportion of mental humility, submission, and independence."

By practicing what he preached, the intellectually irreverent and religiously humble young philosopher earned such a respected position at the Royal Institution, he no longer worried about ever having to go back to bookbinding. Now, he thought cheerfully, rolling up his sleeves, he could concentrate on that other childhood dream of his—the one about becoming the first to demystify the puzzling phenomenon of electricity; unfortunately for Faraday, however, others around the world had grown up having the same dream and now were coming very close to realizing it.

The closest person appeared to be a Danish physicist named Hans Ørsted. In 1820, he discovered that an electric current caused the needle of a magnetic compass to move slightly, as if the electric current itself were behaving like a magnet.

A few months later, in France, the startling news was confirmed in a slightly different way by Ampère and a colleague, Dominique François Jean Arago. They discovered that an electric current in the shape of a corkscrew also behaved like a magnet, attracting iron fillings; for that reason, they called their discovery an *electromagnet*.

Over the past two centuries, natural philosophers had discovered various similarities between electricity and magnetism. The Frenchman Charles-Augustin Coulomb had found that both forces *looked alike;* they weakened with distance in exactly the same way. And the German Otto von Guericke had found that both forces were *two-faced*; they were capable of repelling some objects and attracting others.

Now, Faraday reflected incredulously, Ørsted, Ampère, and Arago had revealed something more, something deeper about the two forces. Their stunning discovery now raised the possibility that electricity and magnetism were somehow interchangeable.

If electricity could behave like a magnet, however, it remained to be seen if the reverse was also true: Could magnetism behave like electricity? Put another way: Was a magnet able to produce electricity? Suddenly, finding an answer to that question became the Holy Grail of nineteenth-century science.

Just as Faraday was about to join the search for the sacred truth about electricity and magnetism, however, he was sidetracked by a young woman named Sarah Barnard. Faraday had met the twenty-three-year-old daughter of a Sandemanian elder in church, and though they liked each other very much, he had hurt her feelings by writing a poem that faulted love for distracting men from their work.

In order to win back her affections, ironically enough, Faraday now was obliged to drop everything he was doing. It proved to be difficult, but as a result of applying to the crisis the very same persistence he had demonstrated in his scientific research, eventually he succeeded: On June 12, 1821, the son of a blacksmith was wed to the daughter of a silversmith.

In lieu of a honeymoon—love having distracted him far too much already—Faraday declared his wishes to spend the time writing an article about the history of electricity and magnetism. His forbearing wife Sarah, fully aware of what she had gotten herself into in the first place, gave her consent.

For the next several months, the newly wedded natural philosopher persevered with inimitable intensity. He read everything he could obtain from both the Royal Institution's own library and his friends abroad. In vintage style, furthermore, Faraday redid every single experiment described in the literature, so that he might verify the various results for himself.

By the end of August, after having pored over thousands of facts and replicated hundreds of experiments, Faraday could not put out of his mind one tiny quirk in Ørsted's experiment. Others had noticed it, but this was something so subtle and seemingly inconsequential that only Faraday's uncanny mind for minutiae could have become so engrossed by it.

In the years ahead, in fact, Faraday would often refer back to this moment as a lesson in the importance of being alert to details: "[Science] teaches us to be neglectful of nothing, not to despise the small beginnings . . . for the small often contains the great in principle, as the great does the small."

The magnetism produced by an electric current, Faraday had noticed, always deflected a compass needle the same way: Imagining the compass as lying on a table and the electric current as flowing from floor to ceiling, the needle always moved slightly *counterclockwise*—never clockwise. Faraday wasn't sure what this signified, but after submitting his article on the history of electricity and magnetism to the *Annals of Philosophy,* he set about to figure it out.

As he concentrated, a mental picture began to take shape that explained Ørsted's original experiment. Just as a warm updraft of air sometimes evolved into a tornado, Faraday speculated, a rising current of electricity might very well produce swirling winds of magnetism, causing any nearby compass needle to rotate a bit.

This was more than a guess and less than a theory, Faraday realized, but there was a way of putting it to the test: If an electric current did indeed produce a magnetic tornado, then its churning winds should be able to whirl any magnetic object round and round *continuously,* not just slightly, as with Ørsted's compass needle. The question was how to make that happen.

After fiddling with equipment day and night for weeks, the answer came to Faraday in early September. First, he took a bar magnet and weighed down one of its ends. That way, when placed in a pool of mercury, the bar magnet floated upright, like a tiny buoy.

Next, he planted a vertical wire at the center of the pool and sent an electric current flowing through it, from bottom to top. As a result, the most remarkable thing happened: The buoy-magnet began wheeling around the wire, as if it were being swept around by an invisible current—an invisible *counterclockwise* current.

With this single experiment, Faraday had landed a formidable one-two punch. He had confirmed his magnetic tornado theory and, in the process, had created the world's first electric motor.

In the years ahead, engineers would refine Faraday's crude contraption, creating electric motors that eventually would outmuscle the steam-powered engines that currently were driving the Industrial Revolution. But even a century hence, when electric motors would come in all shapes and sizes, every single one of them would be impelled to spin by the tornadolike magnetic force-field first recognized by England's working-class wunderkind.

In October 1821, the *Quarterly Journal of Science* published Faraday's discovery in an article with the understated title "On Some New Electromagnetic Motions." The report was translated into dozens of different languages, and soon, scientists worldwide feverishly were fabricating their very own facsimiles of Faraday's fabulous find.

Faraday's fame soared, and so did the height of Voltaic piles: In order to obtain the electricity necessary to run electric motors with any significant power, scientists were forced to build the unwieldy batteries so big and tall they filled entire rooms. Until somebody could invent a more efficient source of electricity, it became apparent, steam-powered engines would continue to run circles around Faraday's new machines.

Although the thirty-year-old Faraday was still earning only a lab assistant's salary, he now had the unmitigated respect and admiration of his Royal Institution colleagues—that is, except for one: Humphry Davy. In recent years, the middle-aged chemist had watched young Faraday's skyrocketing scientific career with a curious mixture of pride and jealousy; now he could contain himself no longer.

The showdown between the once and future kings of chemistry began days after the publication of Faraday's *Journal* article. The young man began hearing rumors accusing him of having plagiarized the idea for the electric motor from William Hyde Wollaston, a manager at the Royal Institution.

Wishing to extirpate the accusation, a flustered Faraday wrote to Wollaston straightway:

> I am bold enough, Sir, to beg the favour of a few minutes' conversation with you on this subject, simply for these reasons—that I can clear myself—that I owe obligations to you—that I respect you—that I am anxious to escape from unfounded impressions against me—and if I have done any wrong that I may apologise for it.

Two days later, the two men met face to face. Yes, Wollaston confirmed, he had been experimenting with equipment similar to Faraday's and, like the young philosopher, had come up with the idea about the swirling nature of an electric current's magnetic force-field. Nevertheless, Wollaston assured Faraday, he had not started the slanderous rumors, nor did he countenance them.

Within the next few weeks, Wollaston's avowed support of Faraday silenced the whispers. But it was the silence of Sir Humphry that most troubled the young man. Now that the crisis was over, Faraday was left wondering why his former benefactor had never come to his defense.

Two years later, Faraday got his answer. Having just discovered how to liquefy chlorine, Faraday allowed Davy to read his article before submitting it for publication. This was in keeping with proper protocol, since Davy was both Faraday's boss at the Royal Institution and now president of the ultra-prestigious Royal Society.

Having himself worked for the better part of two decades to liquefy chlorine, the forty-five-year-old Davy was especially eager for the world to acknowledge his role as Faraday's mentor in this climactic achievement. But he went too far. By the time he was done rewording the paper, Davy made it appear as if it was *he* who had given his young protégé the idea that had led to the discovery.

It put Faraday in an awkward position, for whether he complained or not, he was in jeopardy of suffering another Wollaston-like scandal. This time, therefore, the young man chose to relent with humility. "Though perhaps I regretted losing my subject," Faraday later would explain, "I was too much indebted to him for much previous kindness to think of saying that was mine which he said was his."

Two months later, Faraday was nominated for membership in the Royal Society, the Mt. Olympus of English science. It was a measure of the esteem Faraday's colleagues now had for him; it was also to be the final and dramatic step in Faraday's rapid ascension to a throne that had been held by Davy for more than two decades.

Not only did Davy *not* support Faraday's nomination, he campaigned actively against it. On his lunch hour, the knight with less than shining armor would circulate among his colleagues at the Royal Society, reminding them of the Wollaston affair and encouraging them not to vote for the young usurper.

At one point, Davy even demanded of Faraday that he voluntarily remove his name from nomination. "I replied that I had not put it up," Faraday recalled later, therefore "I could not take it down."

In that case, Davy warned, he himself, as president of the Royal Society, would annul the proposal. According to Faraday: "I replied that I was sure Sir H. Davy would do what he thought was for the good of the Royal Society."

On July 1, in an attempt to save the nomination—and his honor and reputation—Faraday published a detailed retelling of the events surrounding his discovery of the electric motor. Once again Wollaston himself corroborated Faraday's remonstration, and once again Sir Humphry remained silent.

This time, however, Faraday was delighted at Davy's muteness, for it meant that he would not stay the election process, as he had threatened to do. Consequently, on July 8, 1824, members of the Royal Society voted in secret, and the outcome was nearly unanimous: There were many white balls cast in favor of Faraday's induction that day . . . and only one black ball cast against him.

Without a conscious desire to do so, the reluctant young warrior had vanquished the English king of science. Faraday still did venerate Davy's talents as a chemist—and would, for the rest of his life—but he disapproved privately of Davy's treacheries as a colleague. "The greatest of all my great advantages," Faraday later would confide satirically, "is that I had a model to teach me what to avoid."

The following year, in 1825, the newest young Fellow of the Royal Society was promoted to director of the Royal Institution. It was, for Faraday, the crowning achievement of his career. Twelve years earlier, he had come to this stately castle of science as a humble servant; now he had become its newest potentate.

In the lab, the unaffected Faraday now worked harder than ever to find the answer to a question that had intrigued him ever since his discovery of the electric motor. If electricity was able to produce magnetism, then why shouldn't the reverse be true—why shouldn't magnetism be able to produce electricity?

Many scientists had been wondering the very same thing but had failed to find an answer. Not even Ørsted had been successful, even though he had been working day and night to find the logical complement to his original discovery.

On August 29, 1831, Faraday struck paydirt. He began by wrapping a long piece of wire around one segment of an iron doughnut, then did the same around another segment, directly across from the first. If the wires were bandages, it would have appeared as if the doughnut's circular arm had been wounded in two opposing places.

Characteristically, Faraday's game plan was straightforward: He would send an electric current coursing through the first wire bandage, producing a magnetic wind that would swirl through the entire iron doughnut. If that magnetic storm were to produce an electric current through the other wire bandage, then Faraday will have discovered what everyone had been seeking; magnetism will have created electricity.

If it happened, Faraday anticipated, then probably the electric current thus produced would be very small; otherwise, almost certainly, others would have seen it long ago. Consequently, Faraday attached to the second wire bandage a meter that would detect even the tiniest dribble of electric current; with that, he now was ready for anything—or nothing—to happen.

As Faraday electrified the first wire bandage by hooking it up to a Voltaic pile, he glanced hopefully at the electric-current meter. Its needle stirred! "It oscillated," Faraday scribbled hysterically into his lab book, "and settled at last in original position."

For a while, Faraday stared at the needle in stupefaction. Would it move again? After a few minutes of waiting in vain, he gave up. As he disconnected the battery, however, Faraday was astonished to see that there was "again a disturbance of the needle."

For the rest of that night, Faraday kept connecting, then disconnecting the iron doughnut; each time he did that, the needle of his electric-current meter danced spasmotically. Finally, an idea dawned on him, and at that moment he was like the young man who had jumped for joy once before, on a Christmas Eve almost twenty years ago.

The electric current through the first wire bandage was producing a magnetic tornado; that whirlwind, in turn, was causing a second electric current to flow through the other wire bandage—but it was doing so only when the tornado's intensity either *increased* or *decreased*. That explained the jumpy needle: Whenever Faraday connected/disconnected the battery, the magnetic tornado suddenly came on/went off, producing the effect. In between those moments, so long as the magnetic winds swirled steadily through the iron doughnut, nothing happened.

It was like a person who had lived all his life near a lighthouse;

he would sit up and take notice if, one day, the foghorn were to stop making its usual sound. Or, if after having been off for a long period of time, it were to start up again. However, so long as the foghorn kept operating without any change, that person did not react.

For the next several months, Faraday revised and refined his apparatus and, in each case, affirmed his original discovery. In 1831, finally, the Royal Institution's forty-year-old prodigy was able to summarize his historic discovery in a single statement:

Whenever a magnetic force increases or decreases, it produces electricity;
the faster it increases or decreases, the more electricity it produces.

Though his colleagues could find nothing wrong with his momentous finding, they were rather amused at Faraday's decision to express it in English. Ever since the seventeenth century, when Newton had invented the calculus, *mathematics* had become the chosen language of science. (See "Apples and Oranges" and "Between a Rock and a Hard Life.")

Even when written flawlessly, any ordinary language—English, Latin, Greek—could be misunderstood as much as 20 percent of the time. By contrast, mathematics appeared to be the only form of communication with which natural philosophers could hope to describe the world with perfect clarity.

In 1831, therefore, Faraday was an anachronism, one of the few notable exceptions to this popular way of thinking. Not only had he *not* trained himself mathematically—indeed, he was quite illiterate in that respect—he believed his colleagues were being misled by their foolish trust in the figments of the mathematical imagination; only the facts from well-run experiments, clearly stated in plain English, were what mattered.

For the rest of his life, Faraday was adamant in his wishes to express his discovery in a way that ordinary people could understand, remaining faithful to the biblical verse that sixteen years earlier had inspired him to clarify the mystery of electricity and magnetism in the first place: "Since the creation of the world, God's invisible qualities—his eternal power and divine nature—have been clearly seen, being understood from what has been made."

Three long decades would pass before the Sandemanian's quaintness would be superseded by modern conventions. In 1865, a young Scottish physicist, James Clerk Maxwell, would publish his landmark *A Dynamical Theory of the Electromagnetic Field,* in which he would translate Faraday's simply stated discovery into a mathematical equation.

Maxwell would use B to stand for magnetism and E to stand for electricity. Also, he would use $-\partial/\partial t$ to stand for the phrase "the rate of increase or decrease of . . ." and $\nabla \times$ to stand for "the amount of . . ." In that case, Faraday's discovery boiled down to this equation:

$$\nabla \times E = -\partial B/\partial t$$

That is, the amount of electricity produced by magnetism was equal to the rate of increase or decrease of the seminal force. A lot of electricity was produced by a rapidly changing magnetic force, whereas barely a trickle was produced by a slowly changing magnetic force. None at all was produced by a magnetic force that remained constant over time.

Though he had expressed himself in what science considered an inelegant language, Faraday had seen the world through the eyes of a poet—that is, where there had been complexity, he had seen simplicity. Together with Ørsted, he had shown that electricity could beget magnetism, and magnetism could beget electricity, a genetic relationship so incestuous and circular there was none other like it in Nature.

Although electricity and magnetism each could assert itself individually, they were, in fact, inextricably conjoined; when one was present, so was the other. For this reason, science eventually christened these queerly-related forces with a single, hybridized moniker: *electromagnetism*.

With this new way of seeing electricity and magnetism, Faraday and his successors finally had realized part of science's ancient dream of unifying the forces of Nature. It was a minor victory, however, in comparison to their overall failure to consolidate all three forces; science's trinity was not as sublime as Christendom's after all.

Neither, as it turned out, was it as sacred. During the twentieth century, scientists were to discover entirely new forces, beyond the original three, complicating even further their vision of how the natural world was created and how its future was being shaped. In retrospect, in fact, the cosmos would never look as simple again as it had in the days when Faraday first had helped reveal to the world the intimate connection that existed between electricity and magnetism.

Because of Faraday's equation, furthermore, people's *lives* would never be as simple again. The son of a common laborer had discerned and written down a great secret of the natural world, one that would spell the end of the Industrial Revolution and the beginning of the Electrical Age.

EPILOGUE

Nearly a hundred years after having won their independence from England, the Americans had gone to war with each other over the issue of slavery. It became one of the most bitter and violent class struggles in history, but now it was over: On April 9, 1865, at

Appomattox, Virginia, Robert E. Lee surrendered to Ulysses S. Grant, and slavery in the United States was about to end.

Back during the War of 1812, news of its conclusion had traveled so slowly that American and British soldiers had kept fighting each other for a full two weeks after a peace treaty had been signed. Now things were very different: Because of the telegraph, news of Lee's surrender spread throughout the world in an instant.

The telegraph had been operational only since 1844, but already it was bringing peoples of the world closer together by enabling them to communicate at the speed of light. First patented by an American painter named Samuel Finley Morse, the telegraph was a direct result of Ørsted's, Ampère's, and Arago's discovery of the electromagnet.

Whenever a sender pressed a telegraph key, it switched on an electric current that traveled through a wire to the receiver's end, where it powered a small electromagnet. Each time that happened, the electromagnet sucked a thin iron tongue to itself, producing a clicking sound; whenever the sender let go of the key, the electric current stopped, the electromagnet went dead, and the iron tongue sprung back to its normal, straight position.

Morse had developed a code, such that the intermittent clicks produced by his novel device could spell out any letter of the alphabet. Consequently, with practice, a very good telegraph operator was now able to send or receive about 150 letters per minute.

The telegraphs themselves had been developed and touted by a number of people, not only Morse, but they had not attracted much attention until the outbreak of the American Civil War. During that conflict, the telegraph had changed military strategy forever, by facilitating communications between field officers and their commanders back at headquarters.

Now that the war was over, and the telegraph had won so much respect, twenty countries decided to sign a pact agreeing to standardize telegraphic equipment and communications. It was the forerunner of the ITU (International Telegraph Union) and of private companies such as AT&T (American Telephone and Telegraph) and IT&T (International Telephone and Telegraph).

Ørsted, Ampère, and Arago had died before seeing what their discovery had wrought, but their colleague Michael Faraday was still alive, though ailing. He had heard about the American Confederacy's surrender and the historic telegraph pact from his nieces, who were attempting to nurse him and his wife, Sarah, back to health.

Faraday himself had just surrendered his thirty-six-year-long directorship at the Royal Institution. His reign had been unprecedented: Never before had someone from the lowest echelon of English society risen to head the Royal Institution and to earn a living at what others before him had done largely for the intellectual pleasure of it all; henceforth, science would cease being a hobby of the independently wealthy and become a profession of the independently minded.

At seventy-three, the humble servant of science now was living in a small house loaned to him by Queen Victoria. Her Majesty felt warmly and behaved generously toward the aging Faraday, but he had not exploited or reveled in the relationship, as might someone more extravagant than he. Besides, earlier in life, he had discovered that such fine indulgences could lead to trouble.

Back in 1844, for example, Faraday had been suspended as an elder of the Sandemanian church for missing a Sunday worship service—the only time that had ever happened to him during his entire life. He had tried explaining to them that he had been dining with the queen, but the austere-minded church fathers had not considered that an adequate excuse.

Though his strict religious beliefs had not allowed him to become socially elitist, they had allowed him to accept the legion of scientific honors conferred on him by admirers everywhere. All together, over the years, Faraday had received no fewer than a hundred titles and commendations from nearly every major country on the planet.

It wasn't that he valued the approbation: Fulfilling his lifelong wish to become a scientist had been reward enough. Faraday had accepted all those awards out of politeness: "I look upon them as honorary memberships," he once explained, "and not to be refused without something like an insult to the other parties concerned."

Though Faraday had been honored for all kinds of noteworthy accomplishments, his greatest achievement had been the discovery back in 1831 that a changing magnetic force produced electricity. That simple insight had changed the world, because it had given life to dynamos, Prometheus-like devices that were able to produce electricity far more efficiently and prodigiously than Volta's piles.

Dynamos created a constantly changing magnetic force simply by spinning a magnet. So long as the dynamos kept rotating, Faraday's equation warranted, they would produce a steady stream of electricity.

Figuring out how best to spin a magnet had been the key in designing the dynamo. At first, back in the 1830s, engineers had used an *electric motor* to spin the dynamo's magnet; the motor itself was kept spinning by siphoning off some of the electricity produced by the dynamo. In other words, the dynamo fed itself, like a person who always reserved some of his body's energy to grow his own food.

Later on, however, engineers had attached paddles to the dynamo's magnet. At first, the magnetic paddle-wheels had been spun by the force of falling water, creating what came to be called *hydroelectric* power plants.

Others afterward had decided to *boil* the water, with the thought of using the resulting steam to turn the magnetic paddle-wheels; it was such a good idea, in fact, even dynamos well into the twentieth century would be powered with steam, though the source of heat would come from a variety of fuels, including nuclear radiation, wood, coal, oil—even cattle manure!

If the steam was produced at a very high pressure, furthermore, engineers had discovered it made the dynamo spin very rapidly. In keeping with Faraday's equation, faster-spinning magnets produced a faster-changing magnetic force and a larger electric current.

By 1865, dynamos had become powerful enough to run the giant arc lamps atop most lighthouses. In the decades ahead, dynamos would increase in size and power, in order to generate enough electricity to operate Alexander Graham Bell's telephone, Thomas Alva Edison's light bulbs, Guglielmo Marchese Marconi's radio— and a growing army of factory machinery.

Dynamos *electrified* the Industrial Revolution, by replacing clunky and inefficient steam-powered engines with relatively quiet and smooth-running electric motors. With the increasing availability of electricity, moreover, people of all classes eventually were to benefit from labor-saving electric appliances such as the vacuum cleaner, iron, and washing machine.

Wherever they were being built, dynamos were energizing the economies of towns across the world. They were helping to create jobs, products, and consumers to such a large degree, in fact, their combined output soon became a measure of a town's prosperity. In the future, the gross national product of a country would rise and fall in step with its total production of electricity, an astonishing correlation that could not be seen with any other form of energy.

In 1867, just as electricity was in the process of boosting the standard of living for millions of people everywhere, Michael Faraday himself, the human dynamo who had made it all possible, finally was beginning to wind down. "I remain in the house useless as to further exertion," he had written a few years earlier, "excused from all duty, very content and happy in my mind, clothed with kindness by all, and honoured by my Queen."

He had labored for forty-plus years, filling up seven large volumes of detailed laboratory notes; he had turned down the presidency of the Royal Society not once but twice; and he had declined the queen's offer of knighthood. "I must remain plain Michael Faraday to the end," he had explained politely.

To friends who stopped by to reminisce and to inquire about his activities, a frail Faraday remarked that he was now "just waiting." He had done everything he had ever dreamed of doing, and then some; now it was time to rest. On August 25, 1867, Michael Faraday, ever vigilant, died sitting up in his favorite chair.

Queen Victoria had offered Faraday his final honor—to be buried with Isaac Newton and other British luminaries in Westminister Abbey. But predictably, the famous scientist had demurred, opting instead to be given "a plain, simple funeral, attended by none but my own relatives, followed by a gravestone of the most ordinary kind, in the simplest earthly place."

Michael Faraday had died as he had lived, wishing to offend neither his God nor his colleagues. "Now that forty years have elapsed," an aging Faraday had written, looking back at everything he had accomplished, "I still hope . . . that I have not either now or forty years ago been too bold."

In three-quarters of a century, Faraday had gone from being a poor, hard-working errand boy to being a poor, hard-working scientist. No one before had changed science and society in such profound and permanent ways, or has since. For that reason, Michael Faraday—the son of paupers and the confidant of princes—always would be remembered for being in a class by himself.

An Unprofitable Experience

Rudolf Clausius and the Second Law of Thermodynamics

*It's no good crying over spilt milk,
because all the forces of the universe
were bent on spilling it.*
—SOMERSET MAUGHAM

Life had been fairly good to him, the fifty-three-year-old Rudolf Clausius reflected; nevertheless, its normal wear and tear had wearied him physically and emotionally. Even worse, he now faced a crisis exceedingly more calamitous than the nagging pain in his knee and all the other minor injuries he had sustained in life: His wife, Adelheid, was in danger of not surviving the birth of their sixth child.

Smiling bravely at the five children seated anxiously on the couch, he fantasized about turning the clock backward; then again, he interrupted, how lucky he was to be who he was—Prussia's most celebrated physicist. Before him, scientists had begun to understand the complicated behavior of earth, air, and water; but it was Clausius, in 1850, who had first discovered the true nature of *fire,* arguably *the* most mysterious of Aristotle's four terrestrial elements.

He had always been a humble man who ascribed little value to the approbation his achievements had earned him worldwide. But at this particular moment, he very much cherished his privileged position, because it was making it possible for his "Adie" to have the best medical care that money and prestige could procure.

As he stared up toward her room, waiting for the doctor to finish the delivery, his wife's screams shattered his soul the way an enemy bullet had once shattered his knee. Unable to remain standing, he fell into the nearest chair and reached for the wailing three-year-old, youngest of the Clausius children.

When this child had been conceived, back in 1872, life had been so much more pleasant and exciting. In that year, Clausius had brought the family back to his beloved Prussia after having been abroad for a long while. The terrible war with the French having just ended, what better way to commemorate their homecoming and the creation of a German empire, Adie and he had giggled, than with a small creation of their own.

That was a year to be cherished for sure, he mused, hugging the toddler closer to him. But if only he had the power to do so, he would reverse the clock even further, to a time *before* the war, *before* the injury he had suffered while volunteering in the military ambulance corps.

War! How similar to the politics of war were the vicissitudes of life, the great Prussian scientist mused, trying desperately to occupy his mind while waiting for news about his wife. The essence of both were the epic and eternal struggles between right and wrong, life and death, victory and defeat.

Also, Clausius thought wearily, in the final analysis, both life and war seemed a bit pointless. But were they really? he wondered. If some cosmic bookkeeper were able to tally up the outcomes of all the struggles—big and small—that had ever taken place throughout the universe for all time, what answer would he come to?

Presuming the struggles could be quantified somehow—like

Olympic competitions—would the bookkeeper find that, in the end, Right had prevailed over Wrong? Life over Death? Victory over Defeat? Or would he discover that it all had added up to a gigantic, meaningless draw?

Earlier this century, Clausius thought, a great victory had been granted to Napoleon I, to the French people. Back then, *theirs* was the empire, *they* were the rulers of Europe. But look what recently had happened! First Napoleon I and now more recently Napoleon III and his people had been defeated—no, humiliated—by the formidable Prussian army. So far as contemporary French and Prussians were concerned, therefore, the net result of all those battles, all that killing, had been an absolute wash.

Lost in reflection, Clausius suddenly realized that his wife had not screamed for some minutes now. She was a fighter, like the French Communards who had defended Paris to the bitter end against the Prussians, even after the rest of France had surrendered. He hated the French, but he admired such courage.

He hoped and prayed, however, that his wife would be more successful in staving off death than had been the Communard martyrs. He wished this for her sake, of course, but also his own and the children's; even the oldest child, the fourteen-year-old, was still too young to be without his mother.

As the minutes passed, the seemingly interminable waiting began to tell on the children's behavior. They had become increasingly restless and tearful, wanting to know how their mother was doing. In an effort to assuage their fears, Clausius headed upstairs, wondering to himself why things had grown so quiet.

He had barely reached the stairs, however, when suddenly the silence was broken by the welcome cry of a newborn infant. It took Clausius a moment to react, but as soon as he recognized that beautiful sound, he bounded up the stairs with a victorious sense of happiness and relief.

His wife had done it one more time, he marveled, but he already had promised God it would be the last time, if only He would spare Adie's life. They both had wanted to have this child—she especially—but he would no longer jeopardize her life for the sake of having a large family.

As the jubilant man reached the landing, the door to his wife's room opened. The doctor stepped out, but oddly, he was not smiling. He beckoned Clausius and confided in a whispered and exhausted voice that his wife had not survived the ordeal. The baby had been born feet first, the doctor explained, creating a struggle during which Adelheid's overtaxed body had simply given out.

Clausius grabbed for the balustrade to steady himself. At first, he was unable to grasp the full meaning of what he just had been told. Shortly, when he did comprehend, he began to break down, but then just as quickly straightened up again, realizing that the children below were looking his way.

Composing himself, Clausius followed the doctor into his wife's room. It was dark inside, the shades having been pulled down, and it smelled of sweat and blood. The room was quiet, except for the crying of the newest member of the Clausius family, a beautiful baby girl.

Timidly, reverently, Rudolf Clausius walked to the bed upon which his Adie lay, her sheets stained crimson. Her eyes were still open, as if she might still be alive, and her skin was still warm. But the stillness of her body quashed his last glimmer of hope. The doctor had not been mistaken; his beautiful and brave wife of sixteen years had lost her war with death.

How ironic, how cruel and painful, was the timeless struggle between life and death, Clausius lamented bitterly, holding his wife's cooling hand. He had devoted his career to the scientific understanding of heat. But as he felt the heat of life pass from his wife's hand, all he could feel was an overwhelming sense of anger over the apparent craziness of human existence.

From the moment we were conceived, Clausius thought, shaking his head ruefully, we did little else but struggle with each other and with death. We pitied a soldier who was killed in war for having died violently. But, in truth, we all spent our lives in a violent, ultimately futile, struggle for survival.

Anyone who had ever lived had recognized that terrible truth, but Clausius now understood it *better* than anyone who had ever lived, and not just because of his wife's death. Twenty-five years earlier, Clausius's revolutionary theory of heat had enabled him to describe Life and Death not emotionally but *quantitatively,* in unprecedented terms.

Consequently, he had been able to compute the answer to that extraordinary bookkeeping question about Life and Death. At any given moment, his calculations had revealed, more things were being killed in the universe than were being born; Death always outscored Life, which explained why each and every life always came to an end. Always.

The universe as a whole was dying, Clausius had found, its life succumbing inexorably—struggle by struggle—to the forces of death. Indeed, even now, at this moment of his most profound grief, the grim imbalance had been maintained: He had lost a wife and gained a daughter, but in his heart and mind, Clausius understood how and why the great equation of life had taken more than it had given.

VENI

In the whole of the universe, there are only two kinds of processes. *Revocable* processes are those whose consequences can be undone, like purchases that can be returned for a full refund or a motion picture that can be stopped and run in reverse. *Irrevocable* processes are those whose consequences are impossible to reverse, like terrible insults whose harm cannot be repaired or the inescapable ravages of time on our bodies.

By being perfectly reversible, revocable processes are able to go on forever, first forward, then backward, then forward again, then backward again, and so forth, ad infinitum. Indeed, in theory, perpetual-motion machines are powered by revocable mechanisms, analogous to the repetitive up-and-down, up-and-down pedaling of a tireless bicyclist.

By contrast, irrevocable processes are mortal. As they proceed, they deteriorate in some indelible way—like an egg that has been scrambled or a tomato that has rotted. Loosely speaking, such things "age" and always end up dying or being destroyed.

"Life would be infinitely happier," Mark Twain once lamented, "if we could only be born at the age of eighty and gradually approach eighteen." Though that might be true, living was inescapably an *irreversible* process: From the moment a life was conceived, its time on earth always flowed from the past, through the present, and into the future; life never flowed the other way around.

At the same time, the seventeenth-century philosopher Isaac Newton noted with some surprise, the overall character of the universe appeared to be *reversible:* Objects rolled uphill and downhill; pendula swung to and fro; things exploded and imploded; in short, for every natural process that behaved one way, there appeared to be a natural process that behaved in exactly the opposite way. Could it be, therefore, that the universe was a cosmic-size *perpetuum mobile,* destined to exist forever?

For most of the eighteenth century, natural philosophers tended to answer in the affirmative, a scientific conclusion wholly in line with common sense and Judeo-Christian convictions. It was difficult for them to imagine the universe's ever coming to an end; worse, it was blasphemous, considering that the Creator Himself was part of it all, eternal and unchanging.

During the late 1700s, however, philosophers were aghast to

discover that the cosmos was *not* wholly reversible after all: There were several natural processes for which there appeared to be no opposing counterparts, and at least two of them had something to do with *heat*.

First, heat seemed always to flow from hot to cold, never from cold to hot. A pot of cold water placed atop a hot campfire, for example, always heated up. It never happened that the water got colder and the fire hotter; that is, a pot of water put on a fire never froze up.

Second, *friction* always changed motion into heat, never the other way around. Simply by stepping on the brakes of a moving vehicle, for example, one could cause the vehicle to stop and the brakes to heat up. But there was no natural mechanism—no such thing as "unfriction"—by which heat spontaneously changed back into forward motion. If there was, the world would be a strange place indeed; for instance, rocks heated by the sun suddenly would start moving on their own, as if possessed by some invisible, purposeful spirit.

The existence of these naturally irreversible processes implied that, like life itself, the universe was aging, changing from one day to the next in a manner that could never be undone. But in what ways exactly did the two irreversible heat processes "age" the universe? And was this aging process ultimately fatal, or would the universe be able to survive it somehow?

These were scientific questions, of course, but to the extent they involved issues of mortality, they soon became intertwined with our deepest philosophical quandaries about human existence. Eventually, in fact, the subject of heat and its effect on the universe extended to the very heart of our religious beliefs.

One who didn't find this growing confluence between the intellectual and spiritual worlds very heart-warming was a Protestant cleric named Ernst Carl Gottlieb Clausius. A devoutly religious man, he believed that God alone could understand the mysteries of our creation and mortality and that Man's own stubborn efforts to do so were arrogant and foredoomed.

Clausius was known as a strict minister by the people of Köslin, a small town in northern Prussia (now Koszalin, situated in the northwest corner of Poland). He was an unyielding traditionalist who kept God's commandments, especially the one that exhorted believers to "be fruitful and multiply."

As the year 1821 drew to a close, Clausius already had thirteen children, and this wife was pregnant with another. The family's excitement over the imminent birth waxed throughout the Christmas holidays and into the new year; finally, the wondrous event happened. On January 2, 1822, Clausius and his wife became the parents of a new boy, whom they named Rudolf Julius Emmanuel.

In that very same year, in Paris, a young French engineer was giving birth to a new era. After years of dogged effort, Sadi Carnot was putting the finishing touches on his magnum opus, *Reflections on the Motive Force of Heat*, which one day would inspire the newborn Clausius to make discoveries about heat that would change the world forever.

The son of Lazare Carnot, Napoleon I's brilliant minister of war, young Sadi had grown up during the early 1800s, at the height of the French Empire. Having witnessed firsthand its demise at the hands of England, Prussia, Austria, and Russia, however, he now wished to see France recover its strength and dignity by tapping into the power of steam.

Already, Carnot warned, England had used steam engines to mine huge amounts of coal for use in smelting unprecedented quantities of iron, a material essential to the future of any industrialized country. In fact, Carnot observed, so essential had steam engines been in making France's archrival a world leader that to take them away now "would be be to dry up all her sources of wealth, to ruin all on which her prosperity depends, in short, to annihilate that colossal power."

It galled young Carnot that English steam engines were more efficient than French ones: For identical amounts of fuel, English engines invariably produced more work. It was mainly to remedy this humiliating and dangerous disparity that Carnot had dedicated his life to the study of these marvelous machines.

Most steam engines burned wood or coal, Carnot had learned, converting water into steam. The high-pressure steam filled the engine's pistons, pressing them downward. When the steam was released through an exhaust port, the pistons returned to their original positions. The exhausted steam was piped to a cool radiator, where it changed back into water, which flowed to the boiler, and was reconverted into high-pressure steam.

A steam engine repeated these steps many times every second. It was a complex piece of machinery, but its essential effect was simple: We provided it with heat, and it provided us with work—though it usually took a great deal of heat to produce very little work.

During this time, it was widely believed that the work an engine produced was determined solely by the temperature of its boiler; that is, the hotter an engine's boiler, the more steam it produced, the faster and harder its pistons moved, and the more work they generated. It seemed like common sense, but as Carnot would reveal in his historic treatise, it was really nothing but common nonsense.

Back in Köslin, several years after his son Rudolf's birth, the Reverend Clausius was in the final stages of preparing his family to move to the nearby village of Überkmünde, where recently he had been invited to start a private school. With so many mouths to feed, the elder Clausius had accepted eagerly this new pedagogical position. It would bolster his meager ministerial income and provide him with a convenient opportunity to influence his children intellectually as well as spiritually.

Überkmünde was located about one hundred miles southwest of

Köslin, so it took a few days for the Clausius family to complete the journey. When they arrived, they were not disappointed: Their new hometown was located on the coast of the Pomeranian Bay of the Baltic Sea, which made for a lovely setting and a relatively stable climate, the seasonal variations being moderated by the water.

As soon as Rudolf was old enough, he began attending the Reverend Clausius's one-room schoolhouse, right along with his brothers and sisters. He had a cheery disposition, a far-ranging curiosity, and a disinclination to follow in his father's ecclesiastical footsteps.

Young Clausius was curious about the *natural* world. In the summer, he loved to hike along the coast, collecting seashells and basking in the warm sunshine. For a change of scenery, he would climb high up into the Pomeranian forest, collecting rocks and digging tiny fossilized seashells from out of the mountainous strata.

In the classroom, young Clausius was eager to know how shells had come to be embedded in mountains so far away from the ocean, and his father was just as eager to explain. According to the Bible and geologists calling themselves *Neptunists,* the Reverend Clausius taught, God's great flood had killed every creature on earth, except for those aboard Noah's Ark. After the waters had subsided, the creatures' carcasses had been left high and dry, entombed in the mud stirred up by the deluge. That was why, the elder Clausius concluded, ministers like himself all over Europe had hung fossils in the rafters of their churches with the inscription: "Bones of Giants Mentioned in the Scriptures."

The Scriptures were also very specific about the date of the flood, the youngster was told. It happened 4,180 years ago—a number obtained by adding up the ages of persons described in the Old Testament. By using the same technique, the minister explained, Neptunists also had estimated the ages of the earth and sun; both were about 6,000 years old.

It wasn't until Rudolf Clausius had to attend high school in the nearby port city of Stettin that he discovered how entirely possible it was to explain the natural world without any reference whatsoever to the supernatural. It was the teenager's first exposure to a nonclerical education, and it was about to ignite in him a lifelong devotion to the study of heat.

In contrast to the religiously inclined Neptunists, young Clausius learned, there were *secular* geologists called *Uniformitarianists*. Recently, one of their chief proponents, a Briton named Charles Lyell, had written a provocative book titled *The Principles of Geology: being an Attempt to Explain the Former Changes of the Earth's Surface by reference to Causes now in Operation.*

Throughout its history, Lyell asserted, the earth had been changed *continuously* and *gradually* by normal, everyday geologic forces, not suddenly and catastrophically by intermittent outbursts of Divine fury. What's more, he wrote, these geologic forces were powered by an inexhaustible supply of heat coming from the earth's own molten interior, much as the human body was sustained by the heat from its core.

Considering that scientists generally were abandoning their old Newtonian-inspired idea of a perpetual universe, Lyell's vision of an inexhaustible earth was decidedly unfashionable, but it was hugely popular nonetheless among his fellow Uniformitarianists. "Until we habituate ourselves to contemplate the possibility of an indefinite lapse of ages," a strident Lyell scolded, "we shall be in danger of forming most erroneous views in geology."

Young Clausius could hardly believe it. The suggestion that the earth was *not* 6,000 years old was exciting enough, but it was even more so to imagine that beneath his feet, thousands of miles down, at the very center of the earth, there could be a heat engine powerful enough to have sculpted the natural world—the mountains, the ocean basins, everything with which he was so enthralled.

As a result of that epiphany, the young man became increasingly fascinated by heat-driven engines. They had been around since antiquity, he learned, but had not become useful until the early 1700s, with major improvements being made in 1764 by Scottish engineer James Watt. Young Clausius even had the opportunity to see a steam pump in operation.

By 1840, in fact, Rudolf Clausius had seen, learned, and done more as a student in Stettin than he had during his entire life growing up inÜckermünde. Two years earlier, ships powered by steam had crossed the mighty Atlantic Ocean for the first time in history. Now, thanks to his eye-opening high school education, he himself felt liberated from the fetters of the past, just like those ships.

Following his graduation, the eager eighteen-year-old entered the University of Berlin, just as five of his brothers had done before him. He began by taking courses in science and mathematics and was immediately captivated by something he learned from his physics professor, Gustav Magnus.

While lecturing one day, Magnus revealed that he had recently made a surprising discovery about body heat. It appeared to be generated by complex chemical reactions that took place in our bloodstream, Magnus explained, not in our lungs, as scientists had always believed.

At that moment, the feeling washed over young Clausius that it might be fascinating and worthwhile to consider seriously the possibility of devoting himself to the study of heat. *Fascinating,* because of heat's central role in the origins of the natural world and in the life of our own bodies. And *worthwhile,* because though they were still wheezing, noisy contraptions, steam engines had matured dramatically in Clausius's brief lifetime, revolutionizing industry and creating lucrative careers for engineers knowledgeable in the mysteries of heat.

By the time he reached his senior year, in 1843, young Clausius was elated with the way things had been going in his life. He had earned good grades and the respect of his professors and classmates.

Equally important, his free-wheeling interests in the natural sciences had finally organized themselves into a short list of favorite topics, with heat vying for top honors.

Suddenly, however, his *esprit de vivre* was snuffed out by word that his mother had died while giving birth to her eighteenth child. Over the years, each pregnancy had robbed her body of some of the strength it needed to survive; now, horribly, all of her stamina had been used up.

Not wishing to burden his father financially, a grieving young Clausius decided to take on a part-time tutoring job. Furthermore, because most of his older siblings were married and already encumbered with obligations to their immediate families, he volunteered to help raise the younger children; that way, Clausius thought, they would not suffer too greatly from the absence of a warm and caring mother.

Although the additional responsibilities took time away from his coursework at the university, Clausius managed to complete his undergraduate studies in 1844. Immediately thereafter, he began his graduate studies at the University of Halle, about one hundred miles to the southwest of Berlin.

Determined not to renege on his promise to help raise his younger siblings, Clausius decided to stay in Berlin and commute into Halle on horseback. As it was a hard day's ride, Clausius made special arrangements with his professors, whereby he would do as much of the required studies at home and travel to campus only for the most essential lectures.

It was an inefficient way to earn a doctorate, but it did have the advantage of granting Clausius freedom to read and learn at his own pace. He began by pursuing his tentative interest in heat and in no time at all discovered himself warming up to the subject.

The young man was particularly intrigued by scientists and engineers who had discovered ways of making heat behave in *unnatural* ways. The Chinese, for example, had invented a device that forced heat to flow from cold to hot, entirely contrary to its normal tendency; called a *refrigerator,* it used ice and worked on the principle of evaporation.

The details of its operation aside, Clausius learned, its ultimate effect was to force heat to flow from inside a cool box to the relative warmth of the room outside. Consequently the cool box became cooler and the warm room became warmer, something that never would happen naturally.

Young Clausius became especially enthralled with the life and work of Sadi Carnot, who observed that steam engines, too, were essentially devices for making heat behave in an unnatural way. They were the antithesis of friction, Carnot explained, able to do what Nature could not: Steam engines routinely converted heat into movement.

What an uncommon insight into the common engine! Clausius was eager to read more of this man's writings, particularly his little booklet titled *Reflections on the Motive Force of Heat,* which Clausius had learned was Carnot's main work.

For months, the young man searched excitedly through bookstores and libraries everywhere, but he came away empty-handed and in the process discovered why. In 1832, when he was only thirty-six years old, Carnot had contracted cholera. By order of the health officer, therefore, all his personal belongings, including nearly all of his papers, had been burned.

Undaunted, young Clausius gleaned whatever he could of Carnot's work by reading secondary sources and was amazed by what he learned. According to the French engineer, the work produced by a steam engine did *not* depend solely on the temperature of its boiler; it depended on the *difference* between the temperatures of its boiler and its radiator. This simply stated formula was considered such a major revelation, Clausius read, it rated being called *Carnot's Principle.*

In order to operate, a steam engine needed not just heat but a *flow* of heat; that occurred only when there was a *temperature difference* between an engine's hot boiler and cool radiator. "The production of heat is not sufficient to give birth to the impelling power," Carnot had concluded, "it is necessary that there should be cold; without it, the heat would be useless."

In plain language, Carnot was suggesting that a steam engine was like a simple mill wheel. Such a wheel worked by tapping water that flowed naturally from a high place to a low place; similarly, a steam engine worked by tapping heat that flowed naturally from a hot boiler to a relatively cool radiator. The bigger and higher the waterfall (picture Niagara Falls), the more horsepower a mill wheel produced; analogously, the bigger and higher the "heatfall," the more work an engine produced.

Clausius was delighted to discover that Carnot had made still one more, equally surprising discovery. According to Carnot's Principle, an engine whose boiler and radiator temperatures were, say, 160 and 40 degrees Celsius, respectively, should produce 20 billion foot-pounds of work for every ton of coal it burned; theoretically, such an engine could lift a 20 billion-pound weight one foot off the ground—or, equivalently, a one-pound weight 20 billion feet off the ground.

When Carnot had measured the actual output of many different engines, however, he had found that the best English engines produced only one-twentieth of that; French engines were even worse. All engines, in other words, appeared to fall far short of Carnot's theoretical ideal. Why should this be? the young Frenchman had wondered.

The short answer was that Carnot's ideal engine represented a perpetual motion machine. In other words, any hypothetical engine whose efficiency corresponded exactly to the difference between the temperatures of its boiler and its radiator could operate forever: Somehow, in theory, the work it produced could be recycled back into heat, which could then be used to fuel the engine, to produce work, to be recycled back into heat, and so forth, ad infinitum.

Like perpetual motion machines, however, Carnot's ideal engines were impossible to build (although that intimidating observation had never stopped skeptics from trying). The world's engineers—whether British or French—could build only *real-life* engines, which never operated at their full theoretical potential, defined by Carnot's Principle.

Impeccably designed and perfectly maintained though they might be, all real-life steam engines were riddled with inefficiencies of one sort or another. One of the worst, Carnot had discovered, was caused by an engine's parts rubbing against each other. That was not surprising, considering that friction (which turned horsepower into heat) was completely antagonistic to the operation of a steam engine (which turned heat into horsepower).

By 1848, as Clausius pondered everything he had read, he began entertaining fanciful thoughts about the fate of the universe: It was aging, scientists believed, because the heat flowing within it was experiencing various kinds of irreversible changes.

Fine, Clausius thought, but what if we could deploy machines throughout the cosmos to force heat into reversing its natural behavior—refrigerators, for example, to force heat to flow from cold to hot? In that way, would we not be able to reverse the cosmic aging process? At the very least, he wondered, might we not be able to *stop* it, so that the universe would stay the same age forever?

He knew, of course, that such a possibility was far-fetched; we could never produce enough machines to do that. But what if there were machines already out there, made by others, or natural machines made by Nature itself? In that case, Isaac Newton and his contemporaries would have been correct after all: The universe would be a gigantic perpetual motion machine, kept alive eternally by engines that forcibly reversed the aging caused by heat's naturally irreversible behavior.

All these questions exhausted young Clausius, but they also made him feel very much alive, the way one felt after a healthful bout of strenuous physical exercise. Above all, they stimulated him to make an irreversible decision: He wanted to be the first person to find the answers.

VIDI

Scientists have always studied the subject of heat as if their lives depended on it, and it was no exaggeration: Their lives—*all* lives—did depend on heat. As Aristotle once observed: Heat is "the source of life and all of its powers—of nutrition, of sensation, of movement, and of thought."

Aristotle, like Hippocrates before him and Galen afterward, believed that body heat came from an inscrutable fire that burned within us, somewhere within the left ventricle of the heart. According to vivisectionists, that was where the blood looked reddest, which they took to mean hottest.

Two thousand years later, in 1833, a British astronomer named John Herschel speculated that heat powered *all* forms of life on this planet, not just humans. Subsequent experiments proved him right, though they also showed that the vivifying heat came not from within the living entities themselves but from the sun—17 million billion kilowatt-hours' worth *every day!*

That tremendous outpouring of heat was what powered all the plants on earth, their leaves—tiny solar panels—converting sunshine into biomass and physical movement. Plants, in turn, sustained the animals, whose myriad activities brought the hubbub of life to the farthest reaches of the earth.

Solar heat, Herschel ventured, breathed life into even *inanimate* phenomena. When heated, for example, air and water expanded and rose, producing currents in their wake. These restless currents, Herschel concluded, were what ultimately gave rise to the lively and often violent weather so characteristic of the earth.

We had always feared the weather, because it could be the agent of human death and destruction. But Herschel's argument made us come to understand that the so-called natural disasters—the hurricanes, tornadoes, and floods—were the earth's *vital signs,* welcome evidence that our home planet was alive and well.

In the end, all of this suggested a rather startling metaphor for the natural world: The sun was like the furnace of a gigantic steam engine, producing heat that powered the earth and everything on it. So long as that furnace stayed hot, all the engines it powered— from human beings to windmills—would never run out of steam.

By extension, scientists reasoned, other suns might be expected to power other worlds in a similar way. Therefore, they concluded, each and every portion of the universe—macroscopic or microscopic, animate or inanimate—could be thought of as being powered by some kind of heat engine.

Not surprisingly, this perception of heat's importance led many scientists to believe that if only they could understand its irreversible behavior, they might understand at long last the irreversible character of life itself. In the time between Aristotle and Herschel, however, scientists created and discarded four different theories of heat before getting it right, and even then, certain questions— about heat *and* life—have remained unanswered to this day.

Initially, the biggest challenge was just figuring out how to *measure* heat. That led the ancient Greeks to make the first guess, Heat Theory #1: "Heat is what produces the sensation of hotness." That was all; it was a rather trivial theory.

It was also wrong, incapable of explaining even this very simple, paradoxical experiment: If a person put her right hand in cold water, it would feel cold; if she put her left hand in hot water, it would feel hot. So far, no surprises. But now, if she put both hands in warm water, then invariably the right (formerly cold) hand would feel hot and the left (formerly hot) hand would feel cold.

The inconsistency illustrated an intolerable flaw in Heat Theory #1. Consequently, scientists were forced to admit that heat was *not* what produced the sensation of hotness; it was the *flow* of heat that did so. This realization led to Heat Theory #2: "Whenever heat *flows into* our bodies, it produces the sensation of hotness: whenever heat *flows out* of our bodies, it produces the sensation of coldness."

This new theory, together with the age-old observation that heat flowed naturally from hot to cold, was enough to explain the paradoxical experiment. In that case, heat from the warm water *flowed into* the cold hand, making the hand feel hot; conversely, heat *flowed out* of the hot hand into the warm water, making the hand feel cold.

This also explained the phenomenon known as *paradoxical undressing.* When a person fell into cold water, his body immediately responded by redirecting the warmth from the outer skin to the vital organs, in an all-out effort to keep them functioning. Gradually, therefore, the outer skin grew colder and colder, until soon it became even colder than the surrounding water.

At that point, since heat naturally flowed from hot to cold, a tiny amount of warmth from the cold water began to *flow into* the even colder outer skin; that produced a sensation of hotness, which caused the victim to strip off his clothes, hastening his own death!

Heat Theory #2 served scientists so well it reigned supreme until 1592. In that year, the famous Italian scientist Galileo Galilei invented the thermometer—or, as he called it, the *thermoscope.*

The contraption was unwieldy, "a glass vessel about the size of a hen's egg," his assistant described it, "fitted to a tube the width of a straw and about two spans long." It was, in effect, a long-necked bottle, which Galileo then turned upside down, placing the mouth into a bowl of water. "This instrument," the assistant recounted, "he used to investigate degrees of heat and cold."

On cold days, the air inside the bottle would contract, causing a slight sucking action that would draw water up into the bottle's neck. The height of the column was a rough measure of how cold it was outside—the colder the temperature, the higher the column.

The thermoscope was, in retrospect, a backward thermometer. But for scientists of that day, Galileo's gawky gizmo represented a forward-looking way of measuring the effect of heat and the basis of a new theory. To wit, Heat Theory #3: "Heat is what causes the column in a thermometer to change height."

Heat theories #1 and #2 had been founded on the undependable, unpredictable *human* sensation of heat. By contrast, this now was a purely *objective* theory; air and nearly all other fluids expanded as they got hotter by amounts precise enough for scientists to measure with a ruler.

Water was one of the few exceptions; oddly, it expanded when *cooled*. Indeed, this peculiar behavior was the main reason for doubting that a person put into a freezer immediately after dying could ever be brought back to life. When frozen, the water in the body's cells expanded, bursting the cell walls beyond repair.

This exception aside, though, seventeenth-century scientists were excited by Galileo's crude thermoscope and undertook to perfect it. Instead of relying on the contraction and expansion of air, which proved to be a bit fickle, they built thermometers made with alcohol. The main problem was that everybody applied different scales of measurement to their instruments; there was no consistency.

A group of Florentine scientists, under the direction of Grand Duke Ferdinand II, for example, used a temperature scale whose high and low marks corresponded to the hottest and coldest days of the Tuscan year, respectively. Not to be outdone, the culinary-minded French used scales whose high mark corresponded to the temperature of melting butter and low mark to the temperature of a Parisian wine cellar.

The first standardized thermometer was not invented until 1714. In that year, a little-known German physicist named Daniel Gabriel Fahrenheit unveiled an instrument that used mercury sealed inside a tiny glass bulb with a long, thin neck. When heated, the mercury expanded beyond the confines of the bulb, rising up the strawlike neck a distance proportional to the heat applied to it.

Fahrenheit had chosen mercury because it expanded uniformly when exposed to temperatures from about *minus* 40 to *plus* 626 degrees—an extraordinary range. Unfortunately, though, the zero-degree mark of his thermometer corresponded to the temperature of freezing *salt* water, which meant that freezing *pure* water corresponded to 32 on his scale and boiling pure water to 212.

People complained that these numbers were too awkward, so in 1742, a Swedish astronomer named Anders Celsius designed a simpler temperature scale whose 0 corresponded to boiling water and whose 100 corresponded to freezing water. The masses objected to this, as well, however, but were mollified when Celsius reversed the two numbers.

In the years that followed, people in all walks of life found a multiplicity of uses for the fabulous new gadgets. Farmers were able to monitor the temperature of animals and incubating eggs, for instance, and meteorologists were able to monitor air temperature. As a result, regional and national weather services were founded worldwide, each of which began to compile the invaluable temperature records scientists would use later to develop apocalyptic theories of global warming.

Eighteenth-century physicians, too, began using thermometers, though the instruments were still quite clumsy. Typically, patients were asked to breathe on the thermometer or hold it in their hands, and it often took the better part of an hour to produce a reliable reading. (It wouldn't be until 1866 that British physician Thomas Clifford Allbutt would invent the small clinical thermometer with which we are so familiar.)

Despite the shortcomings of these early thermometers, however, they provided scientists with an unprecedented opportunity to measure the intensity of the fabled human fire, the source of life. Much to their amazement and excitement, they discovered the fire was imperturbable: Whatever the season, whatever the weather outside, the interior of the human body appeared to remain stubbornly at about 96° Fahrenheit or about 36° Celsius (an estimate later revised upward).

Scientists were in store for other surprises, too, not all of them pleasant. In fact, one of those unexpected disclosures was about to throw cold water all over their lovely new theory.

The villain, as it were, was a Scottish chemist named Joseph Black who worked at the University of Glasgow. In the late 1750s, Black did something seemingly innocuous: He baked equal quantities of mercury and water in an oven and then checked their temperatures. Much to his amazement, the mercury was much hotter than the water. But how could they possibly have different temperatures, he wondered, when both had been heated by the same oven, for the same amount of time?

Anyone who has ever scalded her mouth biting into a piece of freshly baked apple pie might ask the same question. Invariably, the filling is very much hotter than the crust, even though the two have been baked in one and the same oven.

The same phenomenon is commonly experienced at the beach on days when, though the sand is too hot for bare feet and the water is too cold for swimming, the air is just right for lying out on a towel. The sand, water, and air have completely different temperatures, even though they are all baking under the very same sun.

After some thought, Black came to the conclusion that heat traveled in the form of a weightless, invisible, and indestructible fluid. And, judging from his oven experiment, it appeared to him that different objects had different capacities to absorb and retain this thermal fluid, the way different people had different capacities to consume and hold their liquor.

The behavior of someone with a large alcohol capacity—someone who could "hold" his liquor, so to speak—changed very little even after he had imbibed many drinks. Analogously, Black conjectured, the *temperature* of any material having a large "heat capacity" changed very little even after it had absorbed great quantities of thermal fluid.

Conversely, the behavior of someone with a small alcohol capacity changed dramatically—embarrassingly, in most cases—even after just one drink. By the same token, the temperature of any material having a small heat capacity rose quite a lot even if exposed to just a little bit of heat.

The same analysis applied to the pie and beach. Apple filling and sand were materials having very small heat capacities: It didn't take much heat to make their temperatures rise. At the other extreme, pie crust and air had huge heat capacities: They kept their cool even in the hottest surroundings. Water, always the exception, lay somewhere in between.

Fascinating as it was, Black's innocent little experiment had disastrous implications for thermometers. If identical amounts of heat could produce in different materials completely different temperature readings, then thermometers could no longer be regarded as infallible measures of heat. It was that simple, and that painful; it was the end of Heat Theory #3.

It was enough to make even the most clean-living scientist think about going out and getting drunk. Instead, however, Black and others of his day quickly regrouped, creating yet another theory, this time based on his idea of a thermal fluid, which was thereafter christened *caloric,* from *calor,* the Latin word for heat. Heat Theory #4: "Heat consists of a caloric fluid that is invisible, weightless, and indestructible."

In no time at all, scientists grew to like this new theory very much, because it seemed to explain so many things, even simple ones. For example, a material *expanded* when heated, they imagined, because it swelled up in the process of absorbing caloric fluid, the way a dry sponge expanded when it absorbed water.

Here, too, was an explanation for why objects heated up when rubbed together, a major source of inefficiency in the operation of steam engines. The rubbing, scientists now imagined, liberated caloric fluid that had been stored in the objects all along, the way dust was liberated from an old coat when it was brushed vigorously.

Decades later, Sadi Carnot would become a disciple of this theory. In fact, the imagery of a caloric fluid would be at the very heart of his famous comparison of heat engines with mill wheels, inspiring him to assert that "we may justly compare the motive power of heat with that of a fall of water."

With the new theory, however, there did come one big problem: How were scientists supposed to detect a fluid that was invisible and weightless? Here again, Joseph Black assured himself a place in history, by inventing something called (what else?) a caloric meter, or *calorimeter* for short.

In its essence, Black's device consisted of a well-insulated jar through whose lid was stuck a thermometer. When cremated inside the jar, materials released all their heat, all their caloric fluid, into the air trapped inside, and the resulting increase in temperature was measured directly by the thermometer.

How could a person be sure the thermometer reading was a true measure of the heat released, since thermometers had just recently been discredited? It was because the thermometer in this case always measured the heat liberated into the trapped *air*, which always reacted the same way to the same amount of heat: One unit of heat elicited a certain temperature reading, twice the heat elicited twice the reading, and so forth.

Black's calorimeter was a beautifully clever little device, and it did not take long for scientists to adopt it with the same eagerness they once had adopted Galileo's ill-fated thermoscope. This time, however, they felt confident they were on the right track; this time their theory was going to stand up to the heat.

Surprisingly, for an entire century, experiment after experiment validated their optimism. The most spectacular of them all came in 1775, when the famous French chemist Antoine Lavoisier used a calorimeter to discover just how the mysterious heat source of life was able to maintain a constant temperature within the human body.

By now, scientists had long since replaced Aristotle's belief in a *self-sustaining fire* with explanations that were considerably homier and amusingly irreverent. One Scottish doctor, John Stevenson, had speculated that body heat was the result of our body's *composting* the food we ate. "Man's body, of which he is so vain," Stevenson concluded, "is little better than a smoking dunghill."

Even Benjamin Franklin had ventured a theory. "I imagine that animal heat arises by or from a kind of fermentation in the juices of the body," the great statesman–scientist had written, "in the same manner as heat arises in the liquors preparing for distillation."

Being the sober-minded investigator that he was, Franklin had been careful to put his hypothesis to the test. "The liquor in a distiller's vat," he discovered, "has nearly the same degree of heat with the human body; that is, about 94 or 96."

In a series of landmark calorimeter experiments, Lavoisier compared the heat produced by the burning of powdered charcoal with the body heat naturally produced by birds and guinea pigs. (Mercifully, he did not incinerate the poor creatures; he only incarcerated them in a calorimeter.) He also monitored the amount of air consumed by each and observed the gases they exhaled.

Pound for pound, Lavoisier discovered, both the animals and the burning charcoal consumed identical amounts of air and gave off identical quantities of heat. Was this mere coincidence? he wondered. No, he decided, it must mean that animate beings produced their heat the same way inanimate objects did when incinerated; that is, by simple chemical *combustion*.

In the crudest kind of way, Lavoisier had validated Aristotle's 2,000-year-old assertion: The source of life was indeed some kind of fire. But if it was truly the result of ordinary combustion, Lavoisier reasoned, then the fire of life could not be self-sustaining; like any inferno, it would need constant *feeding,* the fuel being the food we ate. Also, it would need *air,* a steady supply of which came from our breathing. To the French aristocrat-scientist, therefore, the body's combustion chamber was probably situated in the *lungs,* not the heart.

During the course of its growth, an average-size carrot absorbed about 20,000 calories of heat from the sun. (Nutritionists today prefer to use Calories, spelled with a capital "C"; one Calorie equals 1,000 calories.) When someone ate that carrot, Lavoisier imagined, all that heat, all that caloric fluid, was liberated by the person's combustion process. That was how the body maintained its constant temperature, how we humans stayed alive.

With triumphant-sounding explanations like that, the caloric theory and calorimeter continued to prosper into the nineteenth century. But then, in 1814, one Julius Robert Mayer was born in Bavaria, a Germanic kingdom second in size only to Prussia. Though he was to have a pitifully tragic life, his ideas were destined one day to help Rudolf Clausius extinguish Heat Theory #4.

As a young man, Mayer was exposed to two completely antagonistic ways of looking at the world. As a high school student at an evangelical theological seminary, he was given the impression that science did not have all the answers. Later, though, in medical school, he was given the impression that *religion* did not have all the answers.

After his schooling, Mayer was not fully satisfied by either tradition, which meant that neither tradition was fully satisfied with him. This was especially well demonstrated when Mayer announced his theory of how the world came into being; like him, the theory was a curious—some called it scandalous—mishmash of science and religion.

In the beginning, Mayer imagined, the universe had been brought into existence by a single, inexpressibly huge force, which he called the *Ursache,* German for "the Cause." Subsequently, the *Ursache* had split up into smaller, diverse *kräfte* ("forces"), each of which now powered some particular aspect of the universe, be it electric, chemical, thermal, and so forth.

Mayer alienated theologians with his lack of reference to God and scientists with his reference to the supernatural-like *Ursache.* Not surprisingly, therefore, Mayer was rejected when he tried publishing the theory in *Annalen der Physik und Chemie* ("Annals of Physics and Chemistry"), one of Europe's most prestigious scientific journals.

Thereafter, even when Mayer's explanations were more conventional, his reputation for being an oddball prejudiced the reviews his work received from peers. At no time was this more evident than in the winter of 1840, when young Dr. Mayer agreed to serve as physician aboard a Dutch merchant ship bound from Rotterdam to Surabaya, Java.

Like most nineteenth-century doctors, Mayer treated his patients by bleeding them, the theory being that a surfeit of blood was what caused the body to become swollen with illness. At first, Mayer did not notice anything abnormal in the blood he let from the sailors. But as the journey took them closer and closer to the tropics, he noticed their blood becoming redder and redder.

This puzzling phenomenon, he decided, was an unexpected validation of the popular caloric theory of heat and Lavoisier's ideas about biological combustion. Back in the Netherlands, he reasoned, the cold weather had forced the sailors' bodies to generate a lot of heat in order to stay warm. In this progressively warmer climate, however, their bodies' combustion mechanism was able to throttle down. Therefore, less of the air inhaled by the sailors was burned up; more of the air was simply absorbed into their blood, causing it to become redder.

Had it been announced by anyone else, this astonishing discovery would have been hailed by the caloric-theory-loving establishment. But coming from this iconoclastic young Bavarian, the elegant explanation was published with very little reaction or appreciation from his peers.

Disheartened but undefeated, Mayer proceeded to do himself even greater harm by incorporating his very credible explanation of the sailors' reddening blood into the framework of his very incredible *Ursache* theory. The cross-breeding produced a chimeric view of the world that horrified the contemporary mind.

According to Mayer, the one huge seminal force that had split up into many smaller and smaller forces was, to this day, still splintering. The sun's force, for example, was now bifurcating into a luminous force (sunlight) and thermal force (solar heat), both of which were being transformed by plants into a chemical force (food), which itself was being split up in multitudinous ways by the living creatures that consumed it.

Some of the chemical force was being converted by the creatures' internal combustion chambers into a thermal force (body heat) and some by their muscles into a mechanical force (body movement). Some of the chemical force, also, was being converted by the creatures' voice boxes into an acoustic force (sounds) and by their brains into an electric force (neural impulses).

Mayer's grand conclusion? The strengths of all the subordinate forces of today—luminous, thermal, chemical, and others yet unnamed—added up exactly to the strength of the original *Ursache*, from which they all had sprouted. In other words, though things everywhere appeared to be changing ceaselessly, the *total amount of force* in the universe was one of the great constants of life; it never had changed, and it never would change.

It was like saying that a barterer's total wealth remained unchanged, even though the number of her possessions increased steadily. It could happen, if the wealth were being split up into an increasing number of lesser and lesser expensive items.

Years hence, in the hands of Clausius and others, Mayer's fantastic assertions would become one of the most sacred theories in all of science. But this was 1842, and though young Dr. Mayer managed to get it published in a respectable journal, *Annalen der Chemie* ("Annals of Chemistry"), his theory of the natural world was widely snubbed.

Most of his colleagues rejected it, based solely on their wariness of the author's reputation for odd ideas. Those few who did bother to contemplate the theory rejected it for speaking of a thermal force being *transformed* into other forces (e.g., the sun's thermal force being transformed by plants into a chemical force). According to the caloric theory, heat—whether called a force or a fluid or whatever—could not be transformed; that is, it could not be destroyed and then reincarnated as something else. Heat, the caloric disciples chanted, was *indestructible*.

In the years ahead, Mayer's despair worsened. Since most scientists had never even read his paper, they were not able to give him credit, even when they began to publish theories reminiscent of his own. In 1847, for example, the great Hermann Ludwig von Helmholtz, a fellow German, published *Über die Erhaltung der Kraft* ("Concerning the Conservation of Force"). It was hailed as a brilliant piece of work, suggesting the exciting possibility that the combined strengths of all natural forces in the universe never changed; yet not once was Mayer's name even mentioned!

By this time, Mayer had come to the very threshold of a nervous breakdown, and his doctors were threatening to commit him to a mental hospital. Mayer's woes increased further when he was arrested by insurgents during the Revolution of 1848, a violent paroxysm of German nationalism. He was released shortly afterward, but two years later, all the frustration and alienation of his tormented life finally caught up with him: One night, unable to sleep, the thirty-six-year-old pariah climbed out of bed and leapt from the window of his second-story apartment.

Much to his chagrin, however, Mayer did not succeed in killing himself; he was still alive, but why? While his colleagues were still trying to understand the *source* of life, now more than ever, he craved only to understand the *meaning* of life. He cursed fate for his continued suffering, not realizing that in this most tragic year of his discontent, his ideas—his life—were about to be validated by a young Prussian physicist who finally would get to the heart of heat.

VICI

In 1848, in Berlin, Rudolf Julius Emmanuel Clausius was leading as many lives as he had names. He was a high school teacher who was well liked for his friendliness and lucidity. He was a graduate student, contemplative and intense, who was only months away from earning his doctorate. And he was a loving, caring surrogate mother to his four younger siblings.

The only thing he lacked was a wife. Friends and neighbors commented constantly on his eligibility as a bachelor, but twenty-six-year-old Clausius always demurred, explaining that, though he had the desire, he had neither the money nor the time to start a family of his own.

For now, Clausius was wed to his studies. Indeed, the only hot prospect he had in mind was completing his thesis and finding a decent-paying job having something—anything!—to do with heat. Recently the caloric theory had been called into question, and young Clausius was eager to become part of the exciting fracas.

Much of the controversy centered on the work of an amateur scientist named James Joule. The son of a wealthy brewer, Joule had grown up in Manchester, England, enchanted by the English scientist Michael Faraday's remarkable discoveries concerning electricity and magnetism. (See "Class Act.")

Recently Joule himself had made a remarkable discovery: Ordinary electricity always heated up the wire through which it flowed and, in the process, lost some of its impetus. A century from now, people would become used to their electrical appliances warming up, especially toasters, TV sets, and light bulbs. But in Clausius's day, no one was quite sure what to make of Joule's discovery.

An influential Irish scientist named William Thomson, for one, argued publicly that Joule probably had observed little more than an extraordinary example of ordinary friction—electricity rubbing its way along the wire, producing heat as it lost some of its own steam; it was a well-known phenomenon, Thomson reminded everyone, that long ago had been explained by the caloric theory.

His bold public assertions notwithstanding, Thomson secretly had begun to have serious doubts about the vaunted caloric theory. But he dreaded the consequences of admitting it, warning that if scientists stopped believing in the indestructibility of heat, "we meet with innumerable other difficulties . . . and an entire reconstruction of the theory of heat from its foundation."

There was now simply too much riding on the caloric theory to abandon it, including Carnot's Principle, which was based firmly on the concept of a caloric fluid. Scientists and steam-engine designers had come to cherish Carnot's Principle; consequently, Thomson was loath to see it discredited by Joule's discovery. For that reason, in a paper published in 1849, the Irishman avowed stubbornly: "I shall refer to Carnot's fundamental principle, in all that follows, as if its truth were thoroughly established."

To make matters even more rousing for young Clausius, the debate over Heat Theory #4 had become political, ever since "the incident" of 1848. In that year, an aggrieved Mayer had written a blunt letter to Joule, accusing him of hogging all the credit for possibly having found fault with the caloric theory. Now, only a short while later, that rancorous personal exchange had grown into a full-fledged nationalistic feud between British and German scientists.

Clausius himself had not yet made up his mind about the caloric theory, but he was quick to side with his fellow German's complaint against the British brewer's son. As Clausius would argue in coming years, Mayer had published his caloric-busting ideas *before* Joule, and in science, publication dates were what established orders of priority.

In some ways, Clausius was reacting like the meticulous scientist he had become, a stickler for precision and protocol. But in other ways, the young man was reacting like a consummate Prussian, fiercely loyal to the cause of German reunification.

Twelve centuries earlier, the Franks had consolidated Germanic lands into something of an empire. But their wondrous creation had been subjugated by the Holy Roman Empire, which itself had been weakened by the Reformation and finally, at the turn of this century, vanquished by the French.

Now the German people resided in a loose confederation of nation-states, a shattered reflection of the mighty empire it once had been. And worse, Clausius lamented bitterly, Prussia itself, though the mightiest of all the German kingdoms, was little more than a French vassal.

The Revolution of 1848 had managed to call attention to the German people's yearning for unity, but already there was some indication that it would not lead to much. It had produced a parliament in Frankfurt, but just recently Prussia's King Frederick William IV had refused to acknowledge its right to offer him an imperial crown.

In reflecting on the German people's sad situation, Clausius was consoled by the thought that he was about to join the worldwide community of scientists. They were not a perfectly unified people, as evidenced by the rift between Joule and Mayer, but at least they fought their wars with words and numbers, not knives and bullets.

In the spring of 1848, young Clausius was awarded his doctorate in science. Out of financial necessity, he retained his high school teaching job but hoped that soon he would be able to afford to marry some young woman and have children.

For now, he began to mull over everything he had ever read on the subject of heat. At long last, his time for being on the sidelines, for merely acquainting himself with other people's theories, was over.

The newly christened scientist wanted to create a theory of his own, but where to begin? As a boy being taught geology, Clausius had learned that science and religion did not always mix well. Unfortunately, he opined, the caloric theory had now become more like a religion than a science, with wavering disciples such as William Thomson trying hard not to lose their faith. Scientists, he insisted, had to rely on facts, not faith.

He saw in Joule's exacting experiments the *factual* basis and in Mayer's offbeat speculations the *philosophical* basis of a whole new way of thinking about heat. The two simply needed to be woven together, warp and woof, in the loom of mathematics. The task wouldn't take long, the young scientist imagined, but he was wrong: In the end, it took him *eighteen* years to create what was to be the first, and best, intellectual tapestry of his life.

He began this monumental effort in 1850, by publishing a very long paper with a very long title: "On the Motive Power of Heat, and on the Laws which can be Deduced from it for the Theory of Heat." Heat and work, Clausius theorized, were but two varieties of a single phenomenon that came to be called *energy* (a name suggested by the doubting Thomson). Put another way, heat and work were fundamentally the same thing, in that a unit of heat could be exchanged for a unit of work, without affecting the total energy of the universe.

It was as if Clausius were suggesting that rocks and people were two varieties of a single phenomenon called matter. According to this way of thinking, rocks and people were fundamentally the same thing, in that a pound of rocks could be swapped for a pound of flesh, without affecting the total weight of the universe.

Clausius didn't stop there. Just as there were many other varieties of matter, such as leather, wood, metal, and so forth, there were many varieties of this overarching phenomenon called energy. Besides heat (thermal energy) and work (mechanical energy), there were solar energy, electrical energy, and acoustic energy, to name just a few.

According to Clausius, in Joule's enigmatic experiment, electrical energy was being changed into thermal energy; that is, as the wire heated up, the electricity flowing through it slowed down, with exact reciprocity. More generally, a unit of *any* one kind of energy could be changed into a unit of any *other* kind—without affecting the total energy of the universe.

This novel concept came to be called the *Law of Energy Conservation,* according to which energy could not be created or destroyed, merely transformed from one variety to another. The total energy of the universe was a true constant of life, Clausius concluded; the only thing that ever really changed was the *mix* of different kinds of energies.

Using the coded language of mathematics, Clausius's brainstorm could be summarized in far less space than it took to spell out in plain English. Choosing $E_{universe}$ to stand for the total energy of the universe and the capital Greek letter delta, Δ, to stand for "the net change in . . . ," the Law of Energy Conservation boiled down to this mathematical equation:

$$\Delta E_{universe} = 0$$

That is, the net change in the total energy of the universe is always zero, because the total energy of the universe is an eternal constant.

Clausius's reasoning spelled the end of the caloric theory, because it recognized *energy,* not heat, as being the indestructible phenomenon. That unprecedented idea led to Heat Theory #5:

"Heat is but one of many different kinds of energy, all of which can be exchanged for one another at any time, without any net effect on the total energy of the universe."

Upset though they were to see anyone suggest a replacement for their beloved caloric theory, William Thomson and other scientists nevertheless were pleased about one thing. Even if this new theory was adopted, it would not lead to the demise of Carnot's precious rule, only a reinterpretation of it.

According to young Clausius, Carnot's Principle had been correct in saying that an engine's ideal output was determined solely by the *difference* between the temperatures of its boiler and its radiator. But Carnot had *not* been correct to compare heat engines with mill wheels.

Carnot had imagined that, like water driving a mill wheel, caloric fluid driving a steam engine *survived* the process, flowing from the boiler, then in and out of the pistons, and finally ending up in the radiator; there the caloric fluid was reabsorbed into the water and recycled back to the boiler. According to that quaint metaphor, in other words, the caloric fluid was never actually *consumed* in the process of its being tapped for power, merely pushed around, sucked up, spit out, pushed around, sucked up, spit out, and so forth, over and over again.

In the imagery and vocabulary of Clausius's new theory of heat—the heart of which was the Law of Energy Conservation—thermal energy from the boiler was outright *destroyed* and *changed* into mechanical energy. As Clausius put it: "In all cases in which work is produced by the agency of heat, a quantity of heat is consumed which is proportional to the work done."

Any of the boiler's heat that got as far as the radiator, therefore, was heat that had escaped being changed into work by the intervening pistons—heat that had trickled through the engine's walls and radiated away uselessly into the surrounding air. One might say it was *wasted* heat, Clausius explained, heat that had not produced any work, like water spilling ineffectually past a mill wheel.

A similar profligacy appeared to exist in *all* real-life engines, Clausius observed, from windmills to human bodies. For example, of the total amount of aeolian energy that turned a windmill, only some was transformed productively into mechanical energy, used to pump water or grind corn. The remaining part was transformed into thermal energy by the windmill's vanes rubbing against the air or by the axle rubbing against its bearing, heat that ended up being dissipated uselessly into the air.

Similarly, of the total amount of chemical energy (food) that powered a human body, only *some* was converted beneficially into mechanical energy, used by a person to walk up stairs or lift heavy objects; inevitably, part of it was converted wastefully into useless by-products, excreted by the body's imperfect digestive and metabolic systems.

No engine appeared capable of operating flawlessly, of converting 100 percent of its fuel into useful work. Unless one could switch off friction or create perfect thermal insulation, it appeared Carnot had been correct: Real-life engines always would operate well below their ideal, theoretical potential.

Nevertheless, even in their inherent wastefulness, Clausius reiterated, engines always obeyed the Law of Energy Conservation. In the case of an ordinary, wheezing steam engine, for example, the total thermal energy going into the hot boiler was exactly equal to the work (mechanical energy) produced by the pistons *plus* the wasted heat (thermal energy).

Similarly for windmills and human bodies: The total input always equaled the useful *plus* the wasteful outputs. In short, all the myriad energy changes occurring in all the engines of the universe always tallied up so that there was never any net change in the total energy of the universe. Always!

Clausius's youthful imagination had, without question, produced as radical a theory as had ever derailed the study of heat.

Nevertheless, its physical arguments were so precise, its mathematics so persuasive, that scientists could not resist falling under its spell.

Within a short time, therefore, Rudolf Julius Emmanuel Clausius was being praised all over Europe—and so were Joule and the outcast Mayer, whose work had inspired the young scientist. It was a turning point for all three, but especially for Mayer, who in the years ahead was made a member of the world-famous French Academy of Sciences and awarded their prestigious *Prix Poncelet* for a lifetime of outstanding achievement; by the time Mayer died at the age of sixty-four, he was at peace, having received the credit he had so desperately sought as a tormented young man.

Meanwhile, twenty-seven-year-old Clausius himself was honored with an invitation to teach physics at the prestigious Royal Artillery and Engineering School there in Berlin. He began his new appointment in the autumn of 1850 and acquitted himself so impressively that by December, he was made a *privatdozent* (lecturer for hire) at the University of Berlin. This new position allowed him to charge students attending his lectures a small admission fee, thus enriching his hope that soon he might have enough money for marriage and a family.

In 1851, moreover, a stubborn William Thomson finally decided to recant his outspoken belief in the caloric theory and to support Heat Theory #5. He offered credit to "Mr. Joule, of Manchester, [who] expresses very distinctly the consequences . . . which follow from the fact that heat is not a substance." He even deigned to tip his hat to the foreigner Mayer, whose 1842 paper, Thomson conceded, "contains some correct views regarding the mutual convertibility of heat and mechanical effect."

Thomson also paid well-deserved homage to young Clausius, "who by mathematical reasoning . . . [has] arrived at some remarkable conclusions." But the Briton stopped short of actually giving the Prussian credit for helping him to face facts: "I may be allowed to add that . . . it [the reinterpretation of Carnot's Principle] occurred to me before I knew that Clausius had either enunciated or demonstrated the proposition."

Clausius sensed in Thomson's equivocation something of the nationalistic rivalry that continued to overshadow the scientific study of heat. But Clausius chose to remain above petty disputes. In the years ahead, he continued to mind his manners and to work diligently; his good judgment soon was rewarded.

When Clausius was barely thirty-two, he was offered a professorship at the *Ecole Polytechnicum,* a prestigious new university in Zurich. Even though the celebrated young scientist was crestfallen at having to leave his homeland, the new position paid very well, and he was excited about having the chance to do research alongside some of the finest minds in the world. What's more, his siblings were now old enough to care for themselves.

In the years following the young bachelor's arrival in Zurich, it wasn't long before he was able to accumulate a small fortune and find the love of his life, a young woman named Adelheid Rimpau. Though she lived in Zurich, Rimpau was very much a German—to Clausius's delight—born and raised in Braunschweig.

On November 13, 1859, the two were wed. For about a year they lived in Riesenbach, a suburb of Zurich. Clausius was happier now than he had ever been in his life. His beautiful wife was not only stalwart and talented musically, she shared his desire to have many children—she, too, having come from a very large family.

In 1861, they were overjoyed when Adelheid gave birth to a lovely, healthy girl. Soon thereafter, the young family moved several miles farther away from the center of Zurich, where they could afford a large house and a location "where one has fresh air," Clausius enthused, "and a nice view to the lake and to the mountains."

Clausius was now on top of the world, and from there, he was able to discern the fullest implications of his earlier ideas. Unlike the cool, clean air surrounding the alpine mountains of Zurich, however, his conclusions would prove to be anything but refreshing; disquieting was more like it.

He began his dramatic reasoning by recalling the two familiar examples of heat's irreversible behavior. First, heat seemed naturally to flow from hot to cold, never from cold to hot. Second, friction changed mechanical movement into heat; Nature appeared to have no comparable process for changing heat into mechanical movement.

In its essence, Clausius observed, this lopsided behavior of heat represented two different kinds of change. One represented a *temperature* change (thermal energy flowing from hot to cold). The other represented an *energy* change (mechanical energy changing into thermal energy, via friction).

Was an energy change fundamentally different from a temperature change? Clausius wondered. It reminded him of a similar question he had asked years ago when analyzing steam engines, namely: "Is an increment of heat fundamentally different than an increment of work?" Boldly, he recalled, he had proposed that they were not, that they were but two varieties of one thing—increments of *energy;* that assertion had led him to the Law of Energy Conservation.

By analogy, Clausius now decided to propose something just as far-reaching: Energy changes and temperature changes—like those that occurred in the irreversible behavior of heat—were also but two varieties of one thing—changes in *entropy.* "I have intentionally formed the word *entropy* so as to be as similar as possible to the word *energy,"* Clausius explained, "for the two magnitudes . . . are so nearly allied in their physical meanings, that a certain similarity in designation appears to be desirable."

Years before, Clausius had shown that solar energy, fundamentally speaking, was made of the same stuff as electrical energy or acoustic energy or any other kind of energy. Despite their different sources, despite their different appearances and behaviors, all varieties of energy were covertly one and the same thing.

Ultimately, therefore, all could be reckoned against a common ruler. Thus, whether it was solar, electrical, or acoustic, every type of energy could be measured in terms of, say, calories. It was like saying that every type of massive object—stick, stone, or person—could be ranked in terms of pounds or some other common unit of weight.

Now, Clausius was saying, there was an even larger, more comprehensive phenomenon than energy. Entropy, he imagined, encompassed not only all the varieties of energy but also temperature—temperature being defined, as always, by the readings of an ordinary thermometer.

It was if Clausius had been the first to discover that, large as it was, the United States of America was merely part of a much larger continent. Entropy represented a new and mysterious frontier in scientific thought—one that encompassed the vast territories of energy and temperature and who knew what else—and the young Prussian pioneer was only too eager to explore it.

Despite their different appearances and behaviors, Clausius ventured, energy changes and temperature changes could be reckoned against a common ruler. That is, just as with different brands of energy, these different brands of entropy could be added and subtracted.

A flood of questions now came to the intrepid explorer's mind, among them: What exactly was the sum total of all the entropy changes happening in the universe? Did that grand total fluctuate, or was it a constant? In other words, was there a Law of *Entropy* Conservation, to match his Law of Energy Conservation? If that were true, he thought, glowing, what a fine pair of trophies those two laws would make.

But how would he even begin to carry out such a cosmic-size computation? How could he possibly figure out the total entropy of the universe? That would require him to tally up all the energy changes and temperature changes taking place at any given moment!

Undaunted, young Clausius decided to give it a try, by first creating a simple bookkeeping system: All *natural* changes—energy and temperature changes that occurred spontaneously throughout Nature, without coercion—would be treated as *positive* changes in entropy. For example, wherever heat was escaping from warm houses into the relatively cool outdoors or a cup of hot coffee was slowly getting colder—behavior that came naturally to heat—Clausius would say the entropy at those locations was *increasing*.

Conversely, all *unnatural* changes—energy and temperature changes that occurred only when Nature was coerced by some kind of engine—would be treated as *negative* changes in entropy. For example, wherever there were steam engines changing heat into work or refrigerators forcing heat to go from a cold place to a relatively warm place, Clausius would say the entropy at those locations was *decreasing*.

Armed with a way of keeping the books, the young man now needed to add things up. But how? Years ago, he recalled, he had put the notion of energy conservation to the test by adding up the energy changes that occurred within steam engines. Out of curiosity, therefore, he now undertook to proceed the same way with entropy.

In the machinations of ideal engines, to begin with, Clausius found reason to rejoice. According to his arithmetic, there were exactly as many positive entropy changes as there were negative ones; that is, there was no net change in the entropy of the universe!

Clausius was ecstatic: This was the first bit of evidence that indeed there was a second law identical to the first, a Law of Entropy Conservation! When he continued his computations, however, the blessed rapture gave way to a cursed reality.

For all real-life steam engines (which invariably fell far short of realizing the ideal efficiency defined by Carnot's Principle), Clausius's calculations revealed something completely different. The *natural* changes in such engines (heat escaping wastefully from hot boiler to cool radiator and work being changed wastefully into heat by friction) always *exceeded* the one and only *unnatural* change (heat being changed into work by pistons).

In terms of Clausius's simple bookkeeping scheme, this meant that in any ordinary steam engine, the positive changes in entropy always exceeded the negative changes. That is, the operation of such an engine always resulted in a net *increase* in the entropy of the universe.

The terrible turn of events did not stop there: A stunned Clausius reminded himself that these results applied to every conceivable kind of real-life engine, including windmills and human beings. This meant his discovery about entropy was universal. All the positive and negative entropy changes that occurred in all the real-life engines of the universe always combined so as to *increase* entropy. Always!

In the shorthand of mathematics, Clausius chose $S_{universe}$ to stand for the total entropy of the universe; the capital Greek letter delta, Δ, to stand for "the net change in . . . ;" and the symbol > to stand for "is always greater than . . ." Therefore, his startling result boiled down to this equation:

$$\Delta S_{universe} > 0$$

In plain English: "The net change in the total entropy of the universe is always greater than zero." That is, at any given moment, the *Sturm* and *Drang* of existence always left the universe with *more* entropy than it had the moment before, the positive entropy changes always exceeded the negative ones.

Clausius mused wistfully: For a brief while there, he had been fooled into thinking he had discovered a Law of Entropy Conservation. But that was a law that applied only to a perfect universe, a universe full of *ideal* engines—which was to say perpetual motion machines—where things never aged but went on forever. In that hypothetically ideal universe, entropy was a constant of life, just like energy.

Alas, Clausius sighed, ours was not such a universe. Ours was a universe full of *imperfect* engines—whether they were animate and minuscule, like the cells in our body, or inanimate and gigantic, like the swirling galaxies in the heavens. Ours was a universe where energy was conserved but not exploited with sublime efficiency—a universe, furthermore, governed with inequity by a most mysterious Law of Entropy *Non*conservation.

Clausius wasn't completely disappointed, however: Though his two laws did not match, he was overjoyed to discover that this law that revealed the lopsided behavior of entropy also provided the long-sought-after explanation for the lopsided behavior of heat and of life itself; in fact, his new law was the first scientific explanation for why everything in the universe aged and eventually died!

The universe, this remarkable entropy law revealed, was like a casino. Entropy was like money. Engines were like players.

Clausius's Law of Entropy Nonconservation was like saying that a casino's positive money changes always exceeded its negative money changes. In other words, a casino's winnings always exceeded its losses; it always made a profit, which was how it stayed in business. A casino existed at the expense of its players, which meant it could keep winning only so long as its players could keep losing. When they had lost everything, when positive money changes ceased to exist, the casino would shut down forever.

Similarly, Clausius's Law of Entropy Nonconservation meant that, like a casino, the universe existed at the expense of its engines, the human engine included. So long as the universe continued to make a profit, as it were, so long as the positive entropy changes exceeded the negative ones, it would stay in business. The day all its engines lost everything—the day positive entropy changes ceased to exist—the universe would shut down forever.

There was also another way of seeing it. According to Clausius's bookkeeping scheme, positive entropy changes corresponded to *natural* changes, such as heat flowing from hot to cold or friction changing work into heat. Therefore, his law was equivalent to saying that the universe would shut down forever when all its natural changes had ceased to exist—that is, when all its naturally irreversible phenomena had spent themselves completely.

When would that happen? The number of engines and the size of the universe were far too great for Clausius or anyone else to estimate how long our universe would stay in business. However, he was able to imagine what it would *look* like in its final days.

As thermal energy flowed from hot to cold, it would leave hot areas a little cooler and cold areas a little warmer. Ultimately, therefore, there would be no regions of hot and cold: The entire universe would be uniformly lukewarm.

Without hot and cold regions, heat would cease to flow. That meant, according to Carnot's Principle, that engines would cease operating; they could no longer change heat into useful work.

Friction, in the meantime, would turn any remaining work into heat. That heat would continue to flow from hot to cold until it, too, evened itself out to match the uniform lukewarmness of the dying universe.

Clausius's Law of Entropy Nonconservation portrayed a universe rushing headlong toward a moment when the buzz and fury of its trillions of engines would be silenced forever. It portrayed a universe where mortal violence inevitably gave way to eternal quiescence.

In a manner of speaking, Clausius concluded, his new law painted the picture of an extremely tense universe in the process of letting its hair down, seeking a calmer, albeit moribund, existence.

And therein lay the solution to one of science's greatest mysteries: The irreversible behavior of heat—the irreversible nature of life, in general—was merely an indication that the universe had not yet reached its final resting place.

So long as heat poured forth from all the hot spots in the universe—the stars, the cores of planets, the cores of living bodies—so long as the engines of the universe converted that flow of heat into horsepower, the universe would remain alive, uptight, and violent.

But when that moment came when all the hot spots had cooled off, when all the purposeful mechanical energy had been changed to heat, and that heat, too, had cooled off—only at that moment would peace and quiet prevail in all parts of the universe forever.

For Clausius, the end of his eighteen-year effort had been reached, though not in a way he ever would have imagined. In 1850, he had set out merely to fashion a new theory of heat. He had done that, but he also had come upon an inequality in the laws of Nature that revealed a chilling truism about human existence: We inhabited not a nurturing universe that sustained life but a profiteering universe that existed at the *expense* of life.

As a student of science, Clausius was reassured to think that it probably would be billions of years before the universe won away from us everything we held dear—the earth, the heavens, our children. There was, in other words, no immediate cause for alarm.

As a former student of the Reverend Clausius, however, the forty-three-year-old was disquieted by this unprecedented scientific proof that an end *would* come. He could grasp the mortality of the human body: "All flesh shall perish together," Job 34:15 stated, "and man shall turn again to dust." Clausius could even imagine the impermanence of the earth or the life-giving sun or any other individual aspect of the natural world; but this newly discovered law affected *everything*. One day, he concluded glumly, the whole of God's creation would be dead and gone forever.

EPILOGUE

It was a volcanic eruption that people everywhere would remember for the rest of their lives: On August 26, 1883, the picturesque Indonesian island of Krakatoa exploded, killing 36,000 people and causing the air around the globe to quiver uncontrollably.

The monumental eruption blew so much gas and dust into the upper atmosphere, it blocked out the sun, making it look greenish blue. Consequently, for the next three years, the temperature as far away as Europe dropped by 10 percent, imparting to the summers there an autumnlike coolness.

In Bonn, Germany, sixty-one-year-old Rudolf Clausius marveled at the aftereffects of Krakatoa. They were, in his mind, a dramatic illustration of the force, the determination, with which the universe was tumbling toward its ultimate fate of rest and relaxation, like a boulder bounding down a steep mountainside or, as the poet John Keats had written earlier in the century, like "a fragile dewdrop on its perilous way from a tree's summit."

Like all natural disasters, a volcano was nothing but a big engine. It was powered by the heat that flowed from its very own subterranean pool of molten rock. This so-called *magma chamber* was to a volcano what a boiler was to a steam engine or the metabolic process was to a warm-blooded animal.

The power produced by a volcano was enormous. Whereas a human body produced no more than one-half horsepower and a modest-size steam engine produced hundreds of horsepower, Krakatoa's thunderous eruption had produced more than 30 *billion* horsepower—lifting 20 billion cubic meters of ash and debris more than twenty miles into the air, raising fifty-foot-high waves on the ocean's surface, and dousing the lives of 36,000 people!

There were other effects produced by Krakatoa: Some of its underground source of thermal energy had been spent producing a loud sound, acoustic energy. Some of it had been changed into bright light, luminous energy. Some of it, too, had been *wasted:* The heat simply had flowed from the 2,000-degree-Fahrenheit magma chamber to the relatively cool tropical air of the small island paradise that earlier had been Krakatoa.

According to Clausius's old bookkeeping scheme, some of Krakatoa's catastrophic effects had corresponded to positive entropy changes; others had corresponded to negative ones. All together, however, they had combined so as to *increase* the overall entropy of the universe, just as he would have expected.

The aristocratic-looking old professor shook his white-maned head in amazement: In a flash, 36,000 persons and one volcano had lost everything to the cosmic casino. It would take quite an effort to compute the exact wagering that had gone on, so to speak, but the inevitable conclusion was spelled out by the Law of Entropy Nonconservation: The universe had *profited* from the Krakatoa disaster.

Because of Krakatoa, the universe was now one step closer to realizing its eternal retirement of lukewarm tranquility: Thirty-six-thousand-and-one engines had been stilled. Temperature differences had been evened out: The volcano and the bodies of the victims were now a little cooler; the air around them was now a little warmer.

This much of the grim vision of entropy's aging effects on the universe had been the result of Clausius's original discovery fifteen years ago. Only six years ago, however, in 1877, an Austrian physicist named Ludwig Boltzmann had discerned a different way of describing the same thing.

Entropy, Boltzmann had proved mathematically, was a measure of *disorganization.* Therefore, he had concluded, Clausius's Law of Entropy Nonconservation meant that the universe was becoming more *chaotic* as well as more relaxed.

This implied, of course, that the universe must have started out being very tense and very well organized; it was as if, billions of years ago, Something or Someone had built a superbly designed spring-driven clock and had wound it up good and tight. Like that clock, the universe was now in the process of slowly winding down, slowly relaxing, slowly falling apart.

Right now, the universe was still quite well organized, all its parts operating with scientific precision. There were well-defined regions of hot and cold; there were well-defined and well-designed engines producing well-organized mechanical energy that could be put to well-defined purposes.

With time, however, the universe was losing all of those distinguishing features: Temperature regions were blending into one another, and engines were running out of steam, decaying and blending into the surrounding ground. Even the solid ground itself—all solids, in fact, and liquids, too—were gradually disassociating, everything ultimately becoming a hodgepodge of nondescript lukewarm gases.

Boltzmann's chaotic interpretation of entropy only added to its frightful nature, its incomprehensible ruthlessness. Now, more than ever, it was clear that Clausius's Law of Entropy Nonconservation meant that the universe *preyed* on life and lifelike behavior; it was inclined toward death and destruction.

The creation of life was an unnatural act, a temporary undoing of the natural disorder of things. In short, life *defied* the laws of Nature! So how was this apparent defiance of the entropy law possible? How was it possible for life to come into being in a universe governed by a law inimical to life?

Clausius now knew the answer: Like all unnatural behavior, life was the result of some engine whose coercive effects were able to reverse the laws of normal behavior—the way a refrigerator was able to make heat flow from cold to hot. The engine of life— Whatever or Whoever it might be—was something of a mystery, of course, but one thing about it was certain: Inevitably, its machinations involved entropy changes, some positive and some negative.

The newly created offspring itself corresponded to the biggest of the engine's *negative* entropy changes; that is, the chaos of biological chemicals that resulted from combining a woman's egg with a man's sperm ultimately was transformed into a well-ordered organism, thus *diminishing* the disorganization of the universe. As such, life represented an enormous loss, an unprofitable experience, to the cosmic casino.

According to Clausius's unforgiving entropy law, however, the useful *negative* entropy changes produced by the engine of life always must be exceeded by the wasteful *positive* entropy changes. Scientifically speaking, in other words, the creation of a certain measure of life was unavoidably accompanied by an even greater measure of death.

Clausius knew—and felt—all too well what this meant. He and his beloved wife, Adie, had been engines of life. Together they had given birth to two boys and four girls, but in exchange, they had paid a mortal price.

Back in 1875, Clausius had lost a wife and gained a daughter; in the years that followed, furthermore, the newborn had flowered beautifully. Recalled a family friend: "Never have I met a little girl as cheerful, as joyous, with such a bounce in her step as that last child who was never able to rest on her mother's breast."

But the exchange had not been equitable, the exhausted old man thought. He had taken great pleasure in raising the children, just as once before he had enjoyed raising his own motherless brothers and sisters. But despite his having received great quantities of love and companionship from them all, some part of him could never be cheered, some part of him had died with his precious Adie, irretrievably lost to the cosmic casino.

In the battleground of our daily existence, Clausius had discovered, the forces of Death were stronger ultimately than the forces of Life. He was still living, but he had suffered a net and grievous loss. He was a casualty of the inequitable entropy law; only the universe had gained from the exchange.

Two years before, in 1886, Clausius had remarried. Perhaps, the aged professor thought, brushing at his tear-filled eyes with the back of his hand—perhaps it was his feeble way of trying to make up for the loss of his first love and the loss of his own youth and vigor, his way of trying to defy the entropy law.

Deep in his heart and mind, of course, the elderly Clausius realized that such defiance was futile. The Law of Entropy Nonconservation required that life be lived forward, from birth to death. As the young Austrian psychiatrist Sigmund Freud would put it one day: "The goal of all life is death."

To wish for the reverse was to wish for the entropy of the universe to *diminish* with time, which was impossible. One might as well wish for autumn leaves to assemble themselves in neat stacks just as soon as they had fallen from trees or for water to freeze whenever it was heated.

For Clausius, the season of life was coming to an end. Doctors explained that his body had lost its ability to absorb vitamin B_{12}, which resulted in his having pernicious anemia. His body's fire was flickering, as it were, being smothered by a lack of oxygen.

By the summer of 1888, Clausius's illness had produced irreversible changes to his brain and spinal cord: He couldn't remember things and he had trouble walking. Mercifully, on August 24, he died, surrounded by his adoring family and a few close friends.

His colleagues around the world mourned the loss of a great scientist; his students, the loss of a great professor; his children, the loss of a great father. The world had profited from Clausius's long and productive life. And now that this kind and clever engine had been stilled, the greedy universe as a whole had profited from his death.

$$E = m \times c^2$$

Curiosity Killed the Lights

Albert Einstein and the Theory of Special Relativity

*If a little knowledge is dangerous, where
is the man who has so much as to be
out of danger?*
—T. H. HUXLEY

I t was the spring of 1895, and for sixteen-year-old Albert Einstein, this field trip through the Alps in northeast Switzerland was the nearest thing to paradise he could imagine. For the next three days, he would not have to sit in a classroom and listen to some boring lecture; here he and his curiosity were free to roam up and down some of the most spectacular landscape in the world.

He would have preferred being alone, of course, instead of with his classmates from the Swiss cantonal school in Aarau and his geology teacher, Friedrich Mühlberg. He hated being led around like some pack animal, but he consoled himself by tuning out Mühlberg's running commentary and turning his attention and thoughts to wherever he pleased along the way.

On this particular day, Mühlberg had decided to lead the group up to the summit of Mt. Säntis. It was raining lightly when they set out at dawn, but no one complained, because the misty view was so spectacular, silhouetted against the brightening reddish tint of the eastern horizon.

For hours, the small tribe of students struggled their way up the mountain. The rain intensified, but everyone was wearing hiking boots, so they managed to keep their footing. Everyone, that is, except for Einstein. He hadn't paid much attention to the way he had dressed for the trek; consequently, he now found himself slipping and sliding up the steep-faced slope in his street shoes.

Late that morning, the students were well up the 8,000-foot peak when it happened. Curious about some edelweiss growing out of a dark crack in a huge outcropping of rock, young Einstein leaned too far over and lost his balance. As he began tumbling downward, he tried to grab a bush, a boulder—anything!—but to no avail; he was plummeting to his death.

Just below him, classmate Adolf Fisch looked up and instantly apprehended Einstein's peril. Without hesitation, Fisch fumbled with his climbing stick and held it out, just as his ill-shod classmate came hurdling down. Instinctively young Einstein reached for the alpenstock and held on; his fall had been broken.

Oftentimes, a brush with death causes a person to reevaluate the meaning of his life, to become more introspective, even more religious. But not Einstein; at sixteen, he was already so disengaged from the common realities of life, it was hard to imagine his becoming any more introverted.

As for becoming more religious, young Einstein was Jewish by birth but never had believed in a personal God who dwelled in heaven. Instead, he believed in a pantheistic God who dwelled here on earth, in the flowers, the rain—even the slippery rocks of the Swiss Alps. "I believe in a God who reveals Himself in the harmony of all that exists," a middle-aged Einstein would write, "not in a God who concerns Himself with the fate and actions of men."

Despite his close encounter with death, therefore, the youngster remained curious not about the imponderable beauty of some supernatural Kingdom but about the ponderable beauty of the natural world—for him, a heaven on earth. "I have no particular talent," he would say later in life. "I am merely extremely inquisitive."

In particular, the teenager was "extremely inquisitive" about light. Recently a Scottish physicist named James Clerk Maxwell had offered mathematical evidence for a most extraordinary idea— that light consisted of waves, waves made out of *electricity and magnetism*.

These hypothetical undulations were difficult to picture, but the principle could be illustrated by thinking of a woman trying to adjust the position of a large rug by grabbing one edge of it and flicking it with her wrist; invariably she produced a ripple in the rug that traveled across the room.

According to Maxwell, a similar thing happened every time electricity was switched on (the equivalent of flicking the rug): It always produced an invisible ripple of electromagnetism that traveled across space. That ripple, Maxwell had proved mathematically, was precisely what we called a light wave.

For the past few years now, young Einstein had wondered what a ripple of electricity and magnetism actually looked like. One way to find out, he supposed, would be to pull up alongside the ripple and stare at it. But it was merely an intellectual daydream, he had realized, to suppose one could ever travel at 300 million meters per second, the speed of a light wave.

If only it were *sound* waves that interested him. They traveled at a mere 300 meters per second, making it far easier to imagine what would happen if he pulled up alongside them. What *would* happen? The surprising answer, the young man had concluded, was that he would cease hearing the ripples of sound.

For example, if he were to travel away from an orchestra at

exactly the speed of sound, then his ears would be moving *with* the music (like a surfer riding a wave); consequently, the notes themselves would be moving alongside his ears, not *into* them. As he looked back, he would *see* the musicians, but he would not hear their music.

Could this same thing be true of light? If by some miracle he could travel away from the orchestra at the speed of light, young Einstein had speculated, the inevitable conclusion seemed to be that the light waves would travel alongside his eyes, not *into* them. Therefore, when he looked back at the musicians, he would not see them; it would be as if they all had disappeared!

To young Einstein, this had seemed to suggest a universe too supernatural for his tastes, a place where anything—people, planets, galaxies—could appear to be here one moment and gone the next. Later in his life, as he continued to grapple with this nightmarish brain teaser, he would shake his head in frustration and disbelief, saying: "Who would imagine that this simple law [concerning the speed of light] has plunged the conscientiously thoughtful physicist into the greatest intellectual difficulties?"

For now, however, the sixteen-year-old brushed himself off and breathed a sigh of relief. As he started back down the mountain, with his classmates and teacher trudging protectively nearby, Einstein congratulated himself for having escaped unharmed. The danger was over, he thought, though in truth, it had only begun.

In the years to come, Albert Einstein's indomitable curiosity would lead humanity on an intellectual trek exceedingly more perilous than the rain-plagued hike he had just survived. In pursuing the answers he sought concerning light, furthermore, Einstein would not rest until he reached the very summit of scientific knowledge.

It would be a praiseworthy achievement, but the unexpectedly frightful view from the top would leave us teetering on the precarious pinnacle, wondering what to do next. Should we press onward and upward, to loftier peaks still? Or should we seek a way back down again? These, we would come to realize, were questions science alone could not answer.

VENI

Never before the mid-nineteenth century had there been such hope of using the mathematical and experimental techniques of science to understand, at long last, the origins and behavior of ordinary people. Surely, pundits predicted, the immediate future belonged to the *human* sciences.

In 1859, for example, a British naturalist named Charles Robert Darwin published *On the Origin of Species,* in which he refuted the biblical story of Creation. According to Darwin's heretical new theory, all living things, including human beings, had evolved gradually, via a two-step process he called *natural selection;* it was Nature's version of the *artificial selection* that breeders since the New Stone Age, 10,000 years ago, had used to domesticate countless plants and animals.

The first step in natural selection, Darwin explained, occurred when parents gave birth to offspring. Biologically speaking, though they resembled their parents, the progeny were *unique* individuals, possessing a combination of genes like no one else's.

The second step, Darwin theorized, began with the assumption that, at any given moment, in any given region, there always would be far more offspring coming into the world than conceivably could survive. Therefore, the progeny would be forced to compete with Nature and each other for the limited resources; in the ensuing struggle, Darwin concluded, those scions whose unique genetic traits gave them the greatest advantage would prevail and reproduce.

As an example, Darwin cited the moths that lived in and around the foliage of London. As a result of the Industrial Revolution, the city's buildings and trees had become splotched with sooty pollution. At the same time, Darwin had observed, moths born with naturally spotted wings had flourished at the expense of those born with plain-colored wings; the spotted wings were an advantage, Darwin conjectured, because they blended in with the background and escaped being noticed by predators.

Though he believed strongly in his controversial new theory, Darwin himself was timid about defending it in public. That intimidating task fell to his bravest friends and colleagues, most notably the biologist Thomas Henry Huxley—who others came to call Darwin's bulldog—and the philosopher Herbert Spencer.

In the years that followed, Spencer proved to be exceedingly articulate and persuasive, coining the catchy phrase "survival of the fittest" to explain Darwin's complex ideas to the masses. In the process of championing the theory, however, Spencer took certain unwarranted liberties with it, especially as applied to human society.

According to Spencer, as a result of day-to-day competition in society—at home, at work, in sports, and so forth—the least genetically well-endowed people systematically were being weeded out in a process he called *social Darwinism*. Though scientists, including Darwin himself, scoffed at this perversion of a legitimate theory, it became a popular way for unscrupulous entrepreneurs of the Industrial Age to rationalize their exploitation of the poor.

Arguably the most extreme and chilling exemplar of Spencer-like thinking was the German philosopher Friedrich Nietzsche. "We must seek the superman," he wrote, "who will represent the instincts of competition and survival."

Nietzsche laughed scornfully at humility, compassion, and all the other Christian virtues he believed made people weak and servile. "Observe suffering well and use it as a source of pleasure," he advised arrogantly; "destroy the infirm so that our experience will always be marked by the evidence of a superman."

Spencer's and Nietzsche's way of thinking soon led to a full-blown *eugenics* movement that began promoting the application of age-old selective breeding techniques to humans. The man who named and led the movement was an English psychologist named Francis Galton; in 1874, he wrote the tract *English Men: their Nature and Nurture,* after which he dedicated himself to eugenics research and the creation of national breeding programs to produce intellectual and physical superhumans.

Predictably, it was not long afterward that the strangely evolving science of natural selection became an instrument of evil. By the 1870s, eugenics was being used by government leaders to rationalize their nationalism and by hate-mongers to rationalize their bigotries, including anti-Semitism; to many, eugenics provided indisputable scientific proof that Jews were an inferior, loathsome kind of human being.

In 1879, in Ulm, Germany, this rising tide of scientifically abetted prejudice made life less than pleasant for Hermann and Pauline Einstein. Nevertheless, they had no real choice but to stick it out; not only was Hermann's business located there, but Pauline was pregnant with their first child.

Elsewhere in the world, 1879 was turning out to be a historic year not for ignorance and darkness but for creativity and light: In Menlo Park, New Jersey, for example, Thomas Edison was inventing the light bulb, while in Edinburgh, Scotland, James Clerk Maxwell was nearing the end of an extraordinary life, during which he had been the first to discern the true nature of a light wave.

Also in that year, on March 14, the Einsteins gave birth to a boy, whom they named Albert. In little more than two decades, their son's brilliance would burn with the intensity of a trillion light bulbs, irradiating a mountainous frontier that stretched farther than Maxwell or any other scientist had ever ventured; ironically, however, initial indications were that this newborn had been born mentally impaired.

One might say, in fact, that Albert Einstein's initial development was as slow as light was swift—slow to talk, slow to read, slow to learn. In short, he seemed destined for anything but greatness.

His uncle Jakob, however, preferred to believe that Einstein was merely *distracted,* not dull. Whereas most babies would stare at a mobile placed above their crib, his nephew's attention seemed to be fixed on the enchanting mobiles—the mental imagery—of some unarticulated, *inner* world.

One of the few times young Einstein came out of his shell was when, at five years old, he got his first look at a compass, a gift from his father. The taciturn child had been so startled by the needle's mysterious ability to point north that, in his words later on, he "trembled and grew cold."

In the years that followed, Einstein's development became even more unusual and his upbringing even more unorthodox. His parents did not attend synagogue nor keep a kosher house; shortly after Einstein was born, furthermore, they moved to a Catholic suburb of Munich and thereafter enrolled him at the local Catholic school.

The first day of class was traumatic enough for most kindergartners, but for young Einstein it was particularly jarring. At home, he had been allowed his introversion; but now, with its strict rules, this religious institution forced him to participate in, and conform to, the outside world.

"Worst of all," Einstein would say one day, "is when a school is mainly run by fear, power and artificial authority. All it produces is a servile helot." From that point onward, Einstein came to hate discipline. The more his teachers insisted on conformity, the more like an outsider he felt; it was an emotion that would stay with him for most of his life.

For the next five years, young Einstein complained about having to attend this particular school, but at his parents' urging, he persevered. When time came for him to attend the local secondary school, the Luitpold Gymnasium, things did not improve; he despised its rote style of learning and stern teachers.

Unfortunately for Einstein, the feelings of disapproval were mutual. "You will never amount to anything," his Latin teacher scolded him one day. It wasn't that Einstein came across as a failure; he earned decent grades. It was that he gave the distinct impression of being a smart-aleck.

That impression, moreover, was not entirely inaccurate; Einstein had become quite smart—and cocky—by reading books of his own choosing; guided solely by his curiosity, he had learned far more from those books than from his militaristic teachers at school.

During his first year at Luitpold, for example, Einstein cuddled up to *Popular Books on Physical Sciences,* an engaging collection of volumes written by one Aaron Bernstein. As he read through the pages, the youngster was astonished to learn just how very far nineteenth-century science had come in its description of the universe.

For example, scientists had figured out that the earth spun around its polar axis like a figure skater, creating a centrifugal force that would have long since torn the planet apart had it not been counterbalanced by the earth's own gravitational *self-attraction.* Indeed, Bernstein explained, more than two centuries earlier, Isaac Newton had discovered that this epic tug-of-war had caused our planet to take on the shape of an *orange*—slightly flattened at the poles and bulging around the equator.

For the next several years, young Einstein filled his head with Bernstein's wonderful explanations. He was hooked on the enthralling series, the way many today are addicted to soap operas; just as soon as the youngster had finished one volume, he was eager to begin with the next.

In the process, the ten-year-old became familiar with a brilliant scientist named Rudolf Clausius, who recently had died in nearby

Prussia. Because of Clausius's arresting discoveries concerning heat, young Einstein learned, scientists now were hot on the trail of explaining the extraordinary brightness of the sun and primordial history of the earth.

One of those scientists was an Irishman named William Thomson. According to him, Einstein read with fascination, the sun was so brilliant because it was on fire. Long ago, Thomson believed, the earth, too, had been on fire; furthermore, judging from the present rate at which it was losing heat, he figured, the earth must have cooled down enough to become habitable some 100 million years ago.

Thomson's calculation had unnerved Charles Darwin's defenders, Einstein read in youthful wonderment, because 100 million years was not nearly long enough for natural selection to have worked its effect. In order to account for the plants and animals alive on the earth today, Darwin's provocative theory needed *ten* times more years than that.

In the process of reading his way through all the main ideas of contemporary science, young Einstein even came across a discussion concerning magnetism, the phenomenon that had so startled him as a child. Michael Faraday, he learned, had shown that electricity and magnetism were the two fists of a single force—electromagnetism—though the power behind them remained to this day an intriguing mystery.

Concerned that his son was casting himself too far adrift from ordinary society, Hermann Einstein decided one day to visit the Luitpold Gymnasium. He was led into the headmaster's office, where he proceeded to discuss Albert's problems.

The elder Einstein was not an Orthodox Jew, but he believed that at thirteen, a boy became a man. His son was nearing that age, he explained, and should be giving some thought to a career. What, Hermann Einstein inquired politely, did the esteemed headmaster suggest? "It doesn't matter," came the startling reply.

"He will never make a success of anything."

Over the years, apart from his autodidactic readings, Einstein's idiosyncratic world had been shaped by his mother's interest in classical music and uncle Jakob's success as an inventor. As a result of their influence, young Einstein had come to believe that the natural world was like a sublime symphony or a clever invention: It was beautiful and functioned so well simply because all its parts worked in perfect harmony.

That conviction was reaffirmed most dramatically, when in September 1891, young Einstein came across a geometry book at the local bookstore. That "holy geometry book [made an] indescribable impression on me," Einstein would recall later, because it was perfectly and harmoniously logical, just like Nature.

Einstein's curiosity about the amity between mathematics and Nature increased even more when he learned about an intriguing sequence of numbers, called the *Fibonacci series:* 1, 1, 2, 3, 5, 8, 13, 21, 34, 55, 89, and so on. Even though it was not obvious, there was a pattern to these numbers: Each one was the sum of the two numbers before it (e.g., $13 = 8 + 5$).

First concocted in the thirteenth century by an Italian merchant named Leonardo "Fibonacci" da Pisa, the series had been widely regarded as little more than a numerical curiosity. But then, Einstein learned, botanists had discovered that there were surprising coincidences between the *numerical* pattern of the Fibonacci series and the *growth* pattern of many flowering plants.

As they developed, for example, the branches of a common sneezewort forked in exact accordance with the Fibonacci series: First the seedling's main stem forked (1), then one of its secondary stem's forked (1), then simultaneously a secondary and tertiary stem forked (2), then simultaneously three lesser stems forked (3), and so forth.

Furthermore, Einstein learned, the numbers of petals of various flowers, too, recapitulated the numbers of the Fibonacci series: An iris almost always had three petals, a primrose five petals, a ragwort thirteen petals, a daisy thirty-four petals, and a michaelmas daisy either fifty-five or eighty-nine petals.

All these revelations had a single cumulative effect on the young Einstein: Since there was this wonderful parallel between Numbers and Nature, then why not use the laws of mathematics to articulate the laws of Nature? "It should be possible by means of pure deduction," he concluded, "to find the picture—that is, the theory—of every natural process, including those of living organisms."

The beauty of Nature was more than skin deep, he had discovered, and if he wished to describe it artfully, poetically, he would need to labor long and hard to become numerically literate. Therefore, an older Einstein would recall: "Between the ages of twelve and sixteen, I learned the elements of mathematics, including the principles of differential and integral calculus."

During those years, the precocious teenager discovered the secrets of something one might call the *shrinking factor*. It was a mathematical trick he would call upon many years later when struggling to formulate his famous equation.

The shrinking factor, written $1 - s$, referred to any process where the whole of something—a bank account, a tank of gas, a reputation, anything—was shrunk by some small amount s. For example, $1 - 0.01$ meant that the contents of a perfume bottle, say, had been reduced by one-hundredth of its original amount, a mere dab.

The shrinking factor, Einstein learned, could be invoked many times over. In the perfume example, $(1 - 0.01)^5$ was the mathematically concise way of saying the bottle's level had been reduced by a dab every day for *five* straight days. To compute such cases, the young man learned, there was a simple rule, in which N stood for the number of dabs:

$$(1 - s)^N \text{ equaled approximately } 1 - (N \times s)$$

In the perfume bottle example, N was equal to five dabs and s was equal to a hundredth. Consequently:

$$(1 - 0.01)^5 \text{ equaled approximately } 1 - (5 \times 0.01)$$

That was how much perfume was left after five dabs—about 0.95 of the original amount, which was to say about 95 percent.

For any budding mathematician, this was an essential trick of the trade. For Einstein, it was to be the alpenstock that would assist him in hiking up and down the treacherous landscape of his own revolutionary ideas about the natural world.

While Einstein struggled successfully to master mathematics, his father strove *un*successfully to make a go of one business after another. When Albert had been but a year old, his father's engineering workshop in Ulm had failed, which was why the family had moved to Munich. Since then, Einstein's father and uncle Jakob had operated a small electrochemical plant, but now it, too, was going bankrupt. "Most of all," Einstein would reminisce a few years hence, "I have been struck by the misfortune of my poor parents, who for so many years have not had a happy minute."

In the aftermath of this latest fiasco, Einstein's parents and younger sister decided to leave Germany altogether and to journey across the Alps to Italy, where a wealthy branch of his mother's family promised to help them set up a new business. The fifteen-year-old himself was left behind, to live in a boarding house until he finished school; at least, that was the plan.

It took only six months, however, for Einstein and the Luitpold Gymnasium to come to the same conclusion: He had to go. Fed up with Luitpold's authoritarianism, Einstein persuaded the family doctor to write a note excusing him from school, blaming it on "nervous exhaustion." Deciding not to wait for the letter, Luitpold expelled him outright, claiming that "your presence in the class is disruptive and affects the other students."

Several weeks later, when the young refugee showed up in Milan, at his parents' house, they were thunderstruck. Without a diploma, their son would have no chance of garnering any of the best-paying jobs, whether in the military, postal, or railway service. Worse still, even his own growing ambition to be a high school physics teacher might not be realized, because most decent universities would never think of enrolling a high school dropout.

The one notable exception was the famous Federal Institute of Technology (FIT), in nearby Zurich, Switzerland. School regulations allowed any students to attend, so long as they passed the formidable entrance exam. Einstein decided to try but fell victim to his haughty self-confidence.

The teenager ended up doing very well on the mathematics part of the test but scored so poorly on modern languages, zoology, and botany, he failed the exam. It was, he admitted later in life, "entirely my own fault, because I made no attempt whatever to prepare myself."

At this point, Einstein proved to be every bit his father's son. After each business failure, Hermann Einstein had never quit; instead, he always had packed up and gone somewhere else to start anew. Similarly, following his recent failures, Hermann Einstein's teenage son decided to relocate to the picturesque Swiss village of Aarau; there, he would resume his high school studies and prepare for a second try at the FIT entrance exam.

Though his terrible experience at the Luitpold Gymnasium had made him dislike school, Einstein was pleasantly surprised by the Swiss educational system. At Aarau, teachers willingly spared the discipline and spoiled the mind; they indulged Einstein's unrelenting and undisciplined inquisitiveness, and in return, he indulged them with "happy and responsible work such as cannot be achieved by regimentation, however subtle."

He attended the Aarau school for only a year, but in that short time, his private world, which for years had huddled in the shadows of German intolerance, was now suddenly flooded with the light of Swiss sufferance. He felt liberated, energized, ready to explode with unconcealed curiosity. "Sure of himself," a classmate would recall later, "he strode energetically up and down in a rapid, I might almost say crazy, tempo of a restless spirit which carries a whole world in itself."

It was during this time that Einstein came perilously close to losing his life while hiking in the mountains. In straining to count the number of outer petals in an edelweiss some distance away—to see if it was consistent with the Fibonacci series about which he had learned several years earlier—the sixteen-year-old suddenly lost his balance and nearly tumbled hundreds of feet to the ground below.

It was also during this time he began asking questions about the speed of light. Little did the teenager realize that in struggling to find the answers, he would have to confront a scientific establishment no less towering, no less intimidating, than the Swiss Alps themselves.

Undaunted for now, Albert Einstein graduated from high school on September 5, 1896. Full of vigor and optimism, he left Aarau for Zurich, whereupon he retook FIT's entrance exam and, this time, passed it.

Feeling freer than ever to pursue his own curiosity, Einstein proceeded to take advantage of FIT's relatively relaxed environment. He often neglected to do assigned homework, preferring instead to spend long hours reading increasingly technical books of his own choosing, including ones describing Faraday's work on electricity and magnetism and Maxwell's theory of electromagnetic waves, which he considered to be "the most fascinating subject at the time."

In the process, he became increasingly arrogant, expressing contempt for ordinary people and their "philistine" lives. Most of all, he denigrated professors who forced him to do their bidding.

"It is, in fact, a very grave mistake to think that the enjoyment of seeing and searching can be promoted by means of coercion," Einstein would say later. "To the contrary, I believe that it would be possible to rob even a healthy beast of prey of its voraciousness . . . with the aid of a whip, to force the beast to devour continuously, even when not hungry."

Einstein even resented having to take final exams at the end of every semester. "One had to cram all this stuff into one's mind for the examinations," he said, "whether one liked it or not."

"He made no bones about voicing his opinions," an acquaintance would recall later, "whether they offended or not." Unfortunately for Einstein, more often than not, his candid complaints *did* offend others, especially his instructors.

During a field trip, for example, his geology instructor called on Einstein to explain the rock formations they had come upon: "Now, Einstein, how do the strata run here," the teacher asked, "from below upwards or vice versa?" Churlishly Einstein shrugged and said: "It is pretty much the same to me whichever way they run, Professor."

To make matters worse, he also managed to offend his parents back in Milan by falling in love with Mileva Marić, a young Serbian woman of whom they disapproved most strenuously. "You are ruining your future and destroying your opportunities," Einstein's mother pleaded; "no decent family will have her."

Einstein and Marić had met in their freshman year, whereupon he had exulted at having found "a creature who is my equal, and who is as strong and independent as I am!" Next to physics, he professed to love her more than anything in the world, often writing affectionate quartets to her, such as this one:

> Oh my! That Johnnie boy!
> So crazy with desire,
> While thinking of his Dollie,
> His pillow catches fire.

Up until July 27, 1900, the love-struck couple had seemed destined for a life of happiness and success. On that day, however, having completed their coursework and taken the final examination required by the university, they each received their test results.

The letter Einstein received contained wonderful news: He had passed the final exam, earning his diploma. Marić's letter, however, was filled with dreadful tidings: She had failed, having earned good scores on the exam's physics section, but not its mathematics portion.

Adding to the couple's woes, Einstein was forced to pay dearly for his years of insolent independence. He had accumulated what amounted to a 3.3 grade-point average and had every right to expect the FIT faculty to offer him a teaching position; but Einstein received no such invitation. Indeed, certain professors campaigned actively behind the scenes to torpedo any possible job prospects. "I was suddenly abandoned by everyone," Einstein recalled in later years, "a pariah, discounted and little loved."

For the young, would-be scientist, it was a pitiful and hopeless beginning to the new century. By contrast, science itself was entering the next one hundred years full of confidence and high hopes—and for good reason.

Over the past two millennia, science had thoroughly succeeded in solving the essential mysteries inherent in the ancient Greek description of the physical world; as a result, each of the age-old elements Earth, Air, Fire, and Water was now the subject of a thriving scientific discipline. In recent years, science even had managed to tie up two important loose ends concerning the age of the earth and the electromagnetic force.

Just four years ago, in 1896, French scientist Antoine Henri Becquerel had discovered invisible, high-energy emissions coming from uranium ore. Shortly thereafter, husband-and-wife team Pierre and Marie (née Sklodowska) Curie had discovered similar emanations coming from two previously undiscovered elements, which they had named *radium* and *polonium*.

Because these emissions gave every indication of being a *spontaneous* phenomenon—no one had worked to elicit them—it appeared that science had stumbled on a free source of energy. Also, this discovery seemed to have given the embattled Darwinists new life.

Taking into account the heat emanating from these newly found elements in the ground, scientists had refigured their estimates of how rapidly the earth was cooling off. Though still quite speculative, some of their conclusions now indicated the planet *could* have become habitable long enough ago for natural selection to have shaped life.

The other loose end had been tied three years ago, in 1897, when British scientist Joseph John Thomson had discovered a particle tinier even than an atom; it came to be called the *electron* and proved to be the long-sought-after source of Faraday's electromagnetic force; scientists were hopeful that this subatomic particle might also help to explain the inscrutable emissions radiating freely from uranium, radium, and polonium.

In a speech delivered at the turn of the century, the Irish scientist William Thomson congratulated science on having achieved such a marvelous understanding of the natural world. All that remained was a kind of mopping-up operation, he boasted, requiring little more than "adding a few decimal places to results already obtained."

Thomson, however, had neglected to mention the still-unresolved mystery surrounding the ancient Greeks' *fifth* element, the ether, a quintessential substance from which the heavens supposedly had been made. Moreover, he had no idea that looming on the horizon of science was a small dark cloud whose name was Albert Einstein; in just five more years, he would rain all over Thomson's cheery forecast and take by storm science's tidy little description of the cosmos.

VIDI

Light is such an essential part of human existence that the largest part of the brain is devoted exclusively to the interpretation of visual information. More than 60 percent of what we know, cognitive psychologists estimate, is a direct consequence of what we have seen; put another way, if it weren't for the agency of light, we would be 60 percent less enlightened than we are today, putting us about where we were during the so-called Dark Ages.

Most of what we learn through our eyes concerns *space* and *matter,* the two most tangible aspects of reality. Merely by looking, with the aid of telescopes and microscopes, we are able to know the size of the universe and what kind of material it contains.

With our remaining senses, we can fill in the details. In the end, therefore, by noting carefully and systematically its individual sights, sounds, textures, tastes, and smells, we are able to know a great deal about the natural world at large.

Even with the help of all five senses, however, we *Homo sapiens* are unequipped to apprehend clearly *time* and *energy,* the two most *intangible* phenomena in the universe. Unlike space and matter, time and energy are by themselves neither visible nor sensible; indeed, the only way they are knowable to us is by the palpable effects they have on space and matter.

With the passage of time, for example, spatial things tend to change shape—like a leaky balloon that slowly collapses—and material things age. By witnessing these temporal phenomena, we are able to *infer* what time itself must be like.

The same goes for energy. It has the power to transform space and matter in myriad ways, for example, via an explosion; by observing those changes, we are able to acquire an intuitive understanding of what energy itself must be like.

As recently as the late nineteenth century, scientists believed that we would never be able to perceive time and energy independently of space and matter. *Pure* energy and *pure* time, so to speak, were thought to be as imperceptible to us as a pure personality— that is, a personality unattached to a person!

Amazingly, however, despite the severe limitations of our senses, philosophers were able to surmise the behavior of the four phenomena quite well. By the time Einstein was born, in fact, scientists had synthesized a clear-eyed, coherent theory of the universe solely in terms of space, time, matter, and energy.

Opposites though they were, for example, space and time seemed to share at least one very important trait: Both were absolutes, in that everyone, everywhere, reckoned them in exactly the same way. One person's inch was another person's inch; one person's second was another person's second; and so forth.

According to this theory of the cosmos, people never disagreed as to the length, width, or depth of anything spatial or the duration of anything secular. In that respect, the absolute space and time of nineteenth-century science were like universal moral standards, according to which everyone always agreed on what was right and wrong.

That strict moral code also seemed to include speed, defined by the familiar formula:

SPEED = DISTANCE COVERED ÷ TIME ELAPSED

For example, passengers aboard two adjacent trains parked at a station might suddenly be confused if one of the trains started to move forward, ever so slowly and smoothly. Which of them really was moving? the passengers looking through their windows at one another might wonder, unable to feel any rumbling beneath their seats.

Despite their momentary confusion, science believed, those passengers soon would realize which of the trains was moving and which was still parked—if not by some subtle sensory clue (such as being pressed against their seats), then by doing some kind of experiment (such as watching the reaction of marbles rolling around on the floor of the train).

In principle, the motion of one train was *absolutely distinguishable* from the motion of another. In other words, when it came to judging speed, ultimately there were no disagreements. As with space and time, science believed that *speed* was absolute, not relative.

The absoluteness of speed could be illustrated by imagining a spaceship called the *Starlight Express* streaking through outer space. Watching it, let us say, were three tourists, two of whom were aboard ships of their own—one moving *toward* the *Express* at one meter per second, the other moving *away* at that speed. The third person, say, was watching comfortably from the window of a space station parked nearby.

To the space-station tourist, let us suppose, the *Express*'s speed was 100 meters per second (about 200 mph). To the person moving *toward* the ship at 1 meter per second, therefore, the *Express*'s speed was 101 meters per second (the ship's speed *plus* the tourist's own speed). Finally, to the person moving *away* from it, the *Express*'s speed was 99 meters per second (the ship's speed *minus* the tourist's own speed).

According to science's belief in the absoluteness of space and time, the disagreements were illusory. All three tourists agreed on the ship's speed, once they took into account their own different motions with respect to the *Express;* in the long run, that is, they all agreed absolutely that the *Express*'s speed was 100 meters per second.

The same was thought to be true about reckoning the speed of *any* kind of object or phenomenon. If the three tourists were watching starlight, instead of a spaceship, they still would come to the same conclusion; they would all measure slightly different speeds, but after taking into consideration their own speeds, they would all agree absolutely that light traveled at 300 million meters per second.

Not to be left out, the other pair of opposites, matter and energy, also seemed to have at least one thing in common: Both were indestructible; both appeared to obey conservation laws that went something like this: "Matter cannot be created or destroyed, so the total weight of the universe is always the same; likewise, energy cannot be created or destroyed, so the total energy of the universe is always the same." (See "An Unprofitable Experience.")

It might appear as if matter could be destroyed, as when a log was burned and all that remained were ashes. But scientists had come to believe that in such cases, matter was merely transformed, not destroyed; that is, fire changed a log from cellulose to carbon, plus vast quantities of smoky gases, but in the end, the total weight of the combusted materials was the same as the log's original weight.

Similarly for energy. Just as there were different kinds of money—pennies, nickels, dimes—there were different kinds of energy—thermal, acoustic, kinetic, and so forth. And just as it was possible to exchange, say, five pennies for one nickel, Nature constantly allowed one kind of energy to be exchanged for others of equal value.

Kinetic energy, for example, was the energy of *motion*. In the shorthand of mathematics, where m stood for an object's massiveness and v its speed, the formula for it was simple:

$$\text{KINETIC ENERGY} = m \times \tfrac{1}{2} v^2$$

That is, a lightweight, slow-moving object such as a cork floating gently down a river had very little kinetic energy; by contrast, a massive, fast-moving object such as a boulder tumbling down a mountainside had a lot of kinetic energy. (See "Between a Rock and a Hard Life.")

If the boulder were to slam into a tree on its way down, some of its kinetic energy would change into mechanical energy (leveling the tree) and some into acoustic energy (making a loud thwacking sound). With what little kinetic energy it had left over, it would continue tumbling down the mountainside more slowly. The bottom line? In the end, the sum of mechanical, acoustic, and vestigial kinetic energies would equal the boulder's *original* amount of kinetic energy.

Having come up with this well-organized theory of the universe, scientists were then faced with the formidable job of deciding where light fit in. It was a topic that always had baffled them, primarily because light behaved so differently from anything else.

Light appeared to be able to get from here to there *instantaneously,* as if exempt from the laws of ordinary, earthbound existence. Even stranger, its behavior was decidedly ghostlike: Light could pass unscathed through solid, glasslike materials.

For thousands of years, philosophers from Aristotle to Newton had defended the idea that light consisted of tiny particles. Like so many microscopic fireflies, they had reasoned, these specks of light were emitted or reflected by visible objects and received into our eyes; that was allegedly how we saw things.

These imponderable flecks of light supposedly were able to move so sprightly as to appear instantaneous and had no trouble penetrating transparent solids. Furthermore, Newton had explained, these variously sized whits affected the eyes "according to their bigness and mixture—the biggest [are associated] with the strongest colors, reds and yellows; the least with the weakest, blues and violets."

With the weight of Newton's illustrious reputation behind it, this vision of light had come to be treated very seriously, even religiously. On June 13, 1773, however, there was born in London someone who was to cast upon the venerable theory a dark shadow of suspicion.

His name was Thomas Young, and though he preceded Albert

Einstein by more than a century, both men were gripped by a supranatural curiosity about the natural world. Moreover, both were outspoken outsiders, warriors destined to do battle with the scientific establishments of their day.

Ironically, as an infant, Young could not have been more different from Einstein. He was fast to talk, fast to read, and fast to learn; by the time he was sixteen, for example, Young was fluent in nine languages, mathematics among them.

Young went on to become a physician as well as an amateur scientist. At age twenty-six, he dared to suggest that light consisted of waves, not particles, and that "colors of light consist in the different frequencies of vibration."

The most loosely folded waves—ones whose shape resembled gently rolling furrows—caused the eyes to see red. At the other extreme, waves whose shape resembled the tightly folded corrugations in a cardboard box created the impression of violet.

In his own mind, Young tended to compare light waves with ripples on a pond. Whenever two ripples met head-on, he pointed out, instead of colliding, they passed through one another in a ghostlike fashion, just like two light beams; that alone, he figured, was reason enough to disbelieve Newton's particle theory.

In 1799, after having done some brilliant experiments that seemed to prove his point, Young decided to go public. He would take his case to the very heart of the scientific establishment, the Royal Society of London; it was the *sanctum sanctorum* of English science, which counted Isaac Newton himself as one of its most illustrious alumni.

Neither Newton nor his shade could have done a more arrant job of scaring the life out of Young's insurrection, however, than the members who were present that day. One of them, Henry Brougham, was especially imperious: Young's proposed wave theory was "destitute of every species of merit," he scoffed, therefore "we now dismiss . . . the feeble lucubrations of this author, in which we have searched without success for some traces of learning, acuteness, and ingenuity."

It was the first time in his distinguished career that Dr. Young had received such a dressing down. He was humiliated, to say the least, but he remained undeterred.

In the years ahead, he returned to his interest in languages and accomplished many things. At one point, he even became something of a real-life Indiana Jones, helping to decipher the hieroglyphics etched on the famous Rosetta Stone, unearthed in northern Egypt in 1799.

By early the next century, evidence having mounted in opposition to the particle theory, more and more scientists began to embrace the wave theory of light. Some remembered to credit Young, many others did not; nevertheless, when he died on May 10, 1829, Young had the satisfaction of knowing that his earlier humiliation had been avenged at last.

In 1864, furthermore, a persuasive new wrinkle was added to the wave theory: As a result of toying around with the equations of electricity and magnetism, Scottish scientist James Clerk Maxwell found that they predicted the existence of *electromagnetic ripples* that traveled at an astonishing speed—roughly 300 million meters per second.

Coincidentally, that was identical to the speed of light; not exactly instantaneous, as the ancients had once believed, but fast enough to make it seem so. Consequently, then and there, Maxwell leapt to the conclusion that his hypothetical electromagnetic ripples and Young's light waves had to be one and the same thing.

In 1888, Maxwell's mathematical conjecture was confirmed when German physicist Heinrich Hertz used a giant spark generator to produce an effusion of electromagnetic waves. According to the Bible, God had been the first to create light waves from scratch; now Herr Hertz had done it, too.

Throughout the rest of the nineteenth century, the electromagnetic light-wave theory of Young and Maxwell prevailed. It settled many questions about the hitherto puzzling behavior of light, but it also created a new mystery: How were these enigmatic waves able to travel across the perfect vacuum of space, for surely they did, otherwise how could starlight ever manage to reach the earth?

By contrast, sound waves were not able to do that. In a well-known experiment, a clock covered over by an inverted glass bowl could still be heard ticking. When the air was pumped out of the bowl, however, the clock fell silent, its sound waves unable to traverse the surrounding nothingness.

In short, waves needed some kind of material through which to travel: Sound waves traveled through air, ocean waves traveled through water, and rug waves traveled through rugs. So how was it possible for light waves—electromagnetic ripples—to travel across the vacancy of outer space?

Perhaps, scientists speculated, light waves traveled through a material agency that was not easily detectable, some kind of invisible, all-pervasive *ether,* they called it. This ether would be odorless, colorless, and densenessless; and yet, it would enable light waves to convey themselves from one place to another. How convenient!

In 1881, American physicist Albert Michelson and British physicist Edward Morley began a series of extraordinary experiments they hoped would detect the seemingly undetectable ether. It hinged on one idea: Since the earth whirled around the sun at 30,000 meters per second (about 67,500 mph), it would be expected to create quite a measurable wake in the ether, if indeed the invisible stuff really did exist.

Michelson and Morley proposed to compare the speed of light in two different directions—*along* the wake and *across* it. In other words, they would compare a beam of light moving *along* the direction of the earth's orbit with a second beam moving *across* it.

It was as if they were venturing to detect an invisible current of air—the jet stream, say—by observing the speed of an airplane in two different directions. For an airplane flying west to east, the jet stream acted like a tail wind, so the plane's forward speed was measurably *increased*. By contrast, for an airplane flying north to south, the jet stream acted like a cross wind, which deflected the airplane's overall path eastward but left its forward speed measurably *unaffected*.

By applying this same logic to their two beams of light, Michelson and Morley would know there was an ethereal wake—an ethereal jet stream—if one beam appeared to go measurably faster than the other. Otherwise, what else would account for such a discrepancy?

In order to avoid any interference by air currents, Michelson and Morley placed their light source and fancy speedometer inside a tightly-sealed vacuum chamber. Strange as it sounded, scientists believed that even if all the air was removed from a vessel, the ether would be left behind, omniscient and unseen; it could never be eliminated. Consequently, the two scientists reasoned, their apparatus now could be influenced only by an ethereal wake stirred up within the vacuum chamber by the movement of the earth around the sun.

When all these preparations were completed, Michelson and Morley finally ran their experiment, and all went well—except for the results. Much to the scientists' surprise and disappointment, their speedometer had detected absolutely no difference between the speeds of the two light beams.

Their task was full of subtleties, however, and their equipment delicate; therefore, after doing some fine-tuning, the pair of physicists tried once more. But still there was nothing: The speed of light in a vacuum appeared to be exactly the same in both directions!

For the next *twenty* years, Michelson and Morley attempted repeatedly to detect the ether. They tried it by day and by night, and in every season of the year; they fiddled with their apparatus, and tried orienting the light beams every which way, but always, always, the speed of light in a vacuum came out to be the same—300 million meters per second. History's most prodigious effort to detect the ether's ineluctable wake had ended in what now appeared to be the ether's funereal wake.

The enigma of these null results sent scientists all the way back to where they had started: If light consisted of waves and there was no ether, then how was light able to travel across a vacuum? The obvious answer was that the known laws of physics were flawed somehow—either that, or the wave theory of light had to go.

Rather than concede either of those dreadful possibilities, nineteenth-century science turned instinctively to its cherished notions of space, time, matter, and energy. There, scientists declared confidently, they would find a way out of this crisis; instead, however, they unearthed two other dilemmas, both of which called into question their belief in the absoluteness of speed.

In the last century, Michael Faraday had proven that, as if by magic, a moving magnet was able to cause electricity to flow through a nearby wire; amazingly, that simple discovery had spawned the Electrical Age, now in full swing, with Thomas Edison's light bulbs illuminating cities and homes all over the world. (See "Class Act.")

What if the *wire* in Faraday's scenario were moved, instead of the magnet? scientists had wondered. Would the electricity still be produced? Yes, they had discovered; countless experiments had illustrated that electricity was created either way. In other words, the magical effect was always produced, so long as the wire and magnet moved *relative* to one another.

This well-documented behavior of moving magnets and wires created a problem for science, because it was in direct contradiction with its well-known belief that motion was *absolute*, not relative. According to that belief, there was a universe of difference between the magnet moving and the wire moving: Electricity should be produced only when the magnet moved with respect to the wire; absolutely nothing should occur when the wire moved with respect to the magnet.

A second scientific dilemma could be traced back to 1851, when French philosopher Armand Fizeau had discovered that several hypothetical observers who themselves were moving all saw light appearing to move with the same speed. This was *not* how it was supposed to happen.

In the orthodox *Starlight Express* scenario, for example, actual starlight appeared to move at *different* speeds for the different tourists; it was only after factoring in their own individual motions that they were in absolute agreement about the speed of light—a proverbial instance of "All's well that ends well."

In Fizeau's startling experiment, just the opposite was the case. The tourists, as it were, agreed on the speed of light right from the outset, even before they had made their individual adjustments, which meant they were at odds afterward; in the end, in other words, science was left with differences of opinion that it had no way of resolving.

Clearly, the puzzling experiments of Michelson–Morley, Faraday, and Fizeau all added up to imply that something was amiss with science's notion of speed; and since speed was defined as nothing more than "distance divided by time," the experiments also raised the possibility that something was wrong with science's notion of distance and time.

These experimental results, in other words, had the potential of destroying the very foundation of traditional science; yet, as they entered the twentieth century, scientists preferred to think of them as relatively minor glitches that could be resolved easily. They were dead wrong, however, and a cocky, unemployed nobody named Albert Einstein was about to prove it.

VICI

In the summer of 1902, things began looking up for Einstein. His old classmate, Marcel Grossmann, helped to get him a job as a Technical Expert 3rd Class at the Swiss patent office in Bern; in that position, Einstein would be responsible for judging the merits of people's inventions.

It would not be glamorous, but it reminded him of his uncle Jakob, the inventor, from whom he had developed a playful impulse to figure out how things operated. Best of all, every day after work, he would have, as he put it, "the opportunity to think about physics."

Einstein had come a long way from that day when, as a five-year-old, he had been startled by the behavior of a simple compass. In recent years, he had begun to think deeply and critically about magnetism and its alter ego, electricity.

Like other scientists, he was uneasy about the ominous discrepancy between the *relativism* in Faraday's electromagnetic experiment and the *absolutism* in science's views about motion. "The observable phenomenon here depends only on the relative motion of the conductor and the magnet," he noted incredulously, "whereas the customary view draws a sharp distinction between the two cases."

Unlike a scientific insider, the young outcast was disinclined to brush off this disparity. Moreover, he realized that science's belief in absolute motion was the result of its deeper-rooted belief in the absoluteness of space and time; consequently, at stake here was not just some electromagnetic experiment, but the very essence of science's description of the universe.

The longer Einstein thought about it, the closer he came to the conclusion that this glaring discordance was connected somehow to that other puzzling inconsistency—Michelson and Morley's failure to find the alleged ether. Also, he sensed, both were somehow connected to his old childhood fantasy about catching up with a light wave, which was to say, an electromagnetic ripple.

Put another way, Einstein came to believe that science's two unanswered questions "Why does Nature appear to behave in a relativistic way?" and "How do light waves manage to travel across a vacuum?" were related in some way to the somewhat juvenile and whimsical question "Is it possible to catch up with a light wave to see what it really looks like?" The mystery lay in figuring out what that connection might be.

After work each day, the lowly patent clerk applied himself single-mindedly to the task before him. When he needed a break, the young sleuth would go to the Café Bollwerk and bounce ideas off a clutch of friends who called themselves the *Olympia Academy*; there, into the wee hours of the night, they would haggle about the physics of light.

Einstein's only nonscientific diversion during this time was his love affair with Mileva Marić. In January 1902, she and Einstein produced an illegitimate daughter named Lieserl whom they gave away in secret; the world would not learn about this subterfuge until 1986—nor hear from Lieserl herself ever again.

On January 6, 1903, Einstein and Marić finally were married. By August, the young woman was pregnant once more and worried that Einstein might be upset at the prospect of having to support yet another person on his meager clerk's salary. "I'm not the least bit angry that poor Dollie is hatching a new chick," Einstein replied in a note to her. "In fact, I'm happy about it."

Unfortunately, the marriage would not last long, for even though he would father two sons, Einstein's lifelong energies were to be spent giving birth to a scientific revolution, not a family. Indeed, by 1904, he was ready and more than eager to revamp our understanding of the physical universe.

To begin with, in order to be consistent with the relativistic behavior of Faraday's electromagnetic phenomenon, Einstein scrapped the notion of absolute space and time. In his universe, those qualities would be *relative*, in that people would *not* necessarily reckon distance and time in exactly the same way.

According to this new theory, in other words, people would not always agree as to the length, width, or depth of things spatial or the duration of things secular. In that respect, the relative space and time of Einstein's hypothetical universe were like subjective standards: Everyone had a different opinion about what he saw, with absolutely no scientific way of resolving the disagreement.

At this point, a horrified Einstein paused to reconsider where he was heading with this train of thought. It disturbed him to think that the universe could be as anarchic as all that—as nonobjective as a roomful of art critics; certainly, he conjectured, there must be laws to impose reason and order on this imagined chaos of opinions.

After searching for such laws, Einstein finally found them—in, of all places, the Fizeau experiment. According to its puzzling results, the speed of light appeared the same to people moving with different speeds; it was only *after* the people had added or subtracted their own speeds from what they saw that they were left disagreeing irrevocably as to the true speed of light.

It was redolent of that old joke in which a patient complained to her doctor that it always hurt when she flexed her arm. "Then stop flexing your arm!" the wise doctor advised. Similarly, as a way of curing the Fizeau problem, Einstein decided to advise the quarreling observers to stop using the old rules of absolute space and time.

The new rules would be based on the idea, as implied by Fizeau's experiment, that the speed of light appeared the same to *everyone, everywhere.* Though he was fashioning a universe in which space and time were relative, therefore, Einstein was in reality replacing one notion of absoluteness with another one.

By asserting the absoluteness, the inviolable constancy of the speed of light, the young revolutionary was able to deduce the bizarre new ordinances that held sway in his novel universe. It was not easy to accept these new rules, insofar as they defied common sense, but they were relatively easy to understand.

In Einstein's universe, everything changed in principle about any situation involving tourists watching the *Starlight Express* or *any* corporeal phenomenon: They could never agree on speeds, simply because they now had no *absolute* way of even deciding who really was moving or was at rest.

The biggest departure from the orthodox view involved the tourists watching *starlight,* which was to say, electromagnetic waves of any kind. In that case, it was as if the tourists' own motions came to naught, like joggers running in place on imaginary treadmills. Irrespective of their own movements—of the readings on their treadmills' speedometers, as it were—their speed relative to a passing beam of light always appeared to remain the same, 300 million meters per second.

There was another way of seeing this mysterious constancy, Einstein realized. It was as if the different tourists' perceptions of space and time changed in accordance with their individual motions, in such a way that the speed of light—and *only* the speed of light—always appeared the same.

According to this interpretation, Einstein's universe was based on a cosmic-size optical illusion whose confounding effects were universal. No matter how fast a person was moving, his reckoning of an inch and a second always *changed* so as to leave *unchanged* his reckoning of the speed of light!

The effect called to mind Jonathan Swift's most famous traveler. Had Gulliver's own height changed during his strange journey—had he himself *shrunk* while in Lilliput and *grown* while in Brobdingnag—then his impressions about the size of everything and everyone around him would have remained unchanged.

Much to Einstein's delight, such compensatory adjustments

could be described mathematically in terms of a single *shrinking factor*. As a person sped up, he discovered, her perception of an inch and a second *shrank* by a factor that involved only two quantities— her speed v and the constant speed of light c, as measured in the unpolluted vacuum of space.

Expressed in precise terms, it was rather formidable-looking:

EINSTEIN'S SHRINKING FACTOR = $\{1 - v^2/c^2\}^{1/2}$

Nevertheless, it had the telltale form of a basic, garden-variety shrinking factor, $\{1 - s\}^N$. (It was akin to noting that, notwithstanding its details, this phrase had the basic form of a simple sentence—subject, verb, predicate.) Consequently, it was possible for Einstein to simplify it, using the approximation recipe he had learned many years before:

SHRINKING FACTOR equaled approximately $1 - \frac{1}{2} v^2/c^2$

In plain English, for someone at rest ($v = 0$), there was no shrinkage at all; the shrinking factor remained undiminished:

$$1 - \tfrac{1}{2} 0^2/c^2 \ = \ 1 - 0 \ = \ 1$$

For someone moving at a snail's pace—for someone whose speed v was very small—the shrinking factor was reduced by a very small amount, like the perfume in a bottle being reduced by a mere dab:

$$1 - \tfrac{1}{2} v^2/c^2 \ = \ 1 - \textbf{very small amount}$$

For someone moving very fast, however, the shrinking factor was diminished considerably. In short, the faster and faster people moved, the smaller and smaller were their impressions of an inch and a second—surreptitious corrections that resulted in different people, moving with different speeds, always agreeing on the speed of light.

All this raised an important question: "How could Einstein explain Nature's apparent singling out of electromagnetic waves for such special treatment?" Why was it that throughout the whole, vast universe only *their* speed appeared to be absolute?

The answer, Einstein decided, was to be found in the repeated failure of Michelson and Morley and others to find the hypothetical ether. So far as this pragmatic young man was concerned, if there was no *evidence* for an ether, then there was no ether, period.

That dismissal, if correct, meant that electromagnetic waves were able in some mysterious fashion to snake their way through long stretches of thoroughly empty space entirely on their own, void of any material medium; that made them the only waves known to science not connected inextricably with anything *ponderable*. In short, Einstein concluded, electromagnetic-*cum*-light waves were unique in the whole of the universe because they alone represented waves of *pure, massless energy!*

It was no wonder, therefore, that light always had struck philosophers as being so supernatural. Every time one looked at the light from a star, a flame, or even Edison's incandescent bulbs, one was seeing pure, incorporeal energy—as fantastic, in its own way, as beholding a disembodied soul.

For 2,000 years, in one form or another, ether had obfuscated the true cosmos from science's probing senses, but no longer. With his theory of relativity, Einstein had seen the universe through eyes unclouded by the ancient ethereal haze; consequently, that hoary quintessential element was about to become as obsolete as the concept of *absolute* space and time.

As Einstein mulled over his new theory, he discovered it did not affect just space and time. The shrinking factor also applied to that other pair of closely allied quantities, energy and mass—except the other way around. As a person's speed increased, his mass and energy did not shrink, they each *expanded* by the *reciprocal* of the shrinking factor.

In other words, at rest, material objects experienced no change in their normal mass and energy. But whenever they moved slowly, their weight and energy automatically *increased*. As they moved more and more rapidly, furthermore, their heft and energy expanded by leaps and bounds.

What happened, Einstein wondered, when material objects moved as fast as light itself—that is, when v equaled c? In such a case, Einstein noticed, the precise expression (not merely the approximation) of his original shrinking formula was reduced all the way down to zero:

$$\{1 - c^2/c^2\}^{\frac{1}{2}} \; = \; \{1 - 1\}^{\frac{1}{2}} \; = \; 0$$

This meant that for a person traveling at the speed of light, space and time—indeed, the entire visible universe—appeared to *shrink down to nothing*. Reciprocally, furthermore, the person's mass and energy appeared to *expand up to infinity* (the reciprocal of zero being infinity)!

Neither of those really seemed possible, an incredulous Einstein concluded. Therefore, rather than take them seriously, he interpreted those outrageous predictions to mean that his new theory was trying to tell him something, namely that it was physically *impossible* for any material body to travel as fast as an electromagnetic wave—that is, to catch up with a light beam.

At long last, the twenty-five-year-old had stumbled on the answer to the question that had nagged him since he was sixteen: "The years of anxious searching in the dark, with their intense longing, their alternations of confidence and exhaustion, and the final emergence into light—only those who have experienced it can understand."

It was not, however, an answer he had expected or that made him very happy. If his theory were to be believed, then mere mortals would never be able to catch up with an electromagnetic wave on its ceaseless flight, to hold it and turn it every which way, all in hopes of discerning its detailed nature. The most science would ever learn about this extraordinary manifestation of pure energy would come from whatever fleeting glimpses it could obtain from the sidelines, so to speak.

These revelations were radically innovative enough, but the *coup de grace* was yet to come. It happened in 1904, during one of his haggling sessions at the Olympia Academy with a dilettante friend named Michele Besso. "Trying a lot of discussions with him, I could suddenly comprehend the matter," Einstein recalled later. "After my recognition of this, the present Theory of Special Relativity was completed."

What Einstein recognized was this: Scientists were wrong to continue thinking of mass and energy as being phenomena that, though allied, were organically *different*—the way we might think of the two sexes. Science already knew that mass and energy were both indestructible, satisfying identical conservation laws; and now, Einstein had discovered, both behaved exactly alike—that is, both expanded and shrank by an identical factors. In every important respect, Einstein concluded, mass and energy were *indistinguishable* and *interchangeable*. They were like a single person wearing different clothes or sporting different hairstyles; in short, they appeared to be organically identical.

In some ways, this androgynouslike view of mass and energy was reminiscent of science's recent discovery about the close connection between electricity and magnetism. In both instances, then and now, science's picture of the world had become more unified but also more ambiguous and, therefore, less intuitive.

It helped to clarify things somewhat to think of energy and mass as being like, say, U.S. dollars and British pounds. Though dollars and pounds looked very different, they were both essentially the very same thing, monetary forms of exchange. Although the two currencies had different values, furthermore, there existed a rate of exchange, a formula that defined the relationship between them. By analogy, then, the question now facing young Einstein was this: What was the exchange-rate formula that related mass and energy?

The answer, he discovered, could be obtained by climbing aboard the *Starlight Express* one last time. The only thing he had to bear in mind during this final, madcap ride was that, according to his theory, the *Express*'s mass would increase/decrease as its speed increased/decreased.

Quite simply, therefore, if the *Express* were to slow down, then its mass—let it be represented by the letter *M*—would *decrease* by an amount given by Einstein's familiar shrinking factor:

$$1 - \tfrac{1}{2} \, v^2/c^2$$

This factor, Einstein reminded himself, was merely a mathematical way of saying that some whole quantity was being diminished by some fraction, $\tfrac{1}{2} \, v^2/c^2$, of that whole. If we were imagaining an eight-ounce bottle of cologne being diminished by the fraction $\tfrac{1}{4}$, then the loss would amount to $8 \times \tfrac{1}{4}$ ounces; that is, two ounces.

In this case, it was the *Express*'s mass, *M,* that was being diminished as a result of its being slowed down—diminished by the fraction $\tfrac{1}{2} \, v^2/c^2$. Consequently, in analogy with the cologne, the mass loss amounted to $M \times \tfrac{1}{2} \, v^2/c^2$.

As soon as he had written this down, Einstein's keen eye noticed the similarity between it and the well-known formula for *kinetic energy* (energy of motion), which he had learned as a youngster:

$$\textbf{KINETIC ENERGY} = M \times \tfrac{1}{2} \, v^2$$

The *Express*'s mass loss was mathematically equivalent to that much kinetic energy *divided by* c^2:

$$\textbf{KINETIC ENERGY}/c^2 \ = \ \textbf{M} \times \tfrac{1}{2} \ v^2/c^2 \ = \ \textbf{MASS LOSS}$$

In essence:

$$\textbf{ENERGY}/c^2 = \textbf{MASS}$$

For the same reason it was correct to say that if $6/2 = 3$ then $6 = 3 \times 2$, it followed that:

$$\textbf{ENERGY} = \textbf{MASS} \times c^2$$

In the shorthand of mathematics, where E stood for energy and m stood for mass:

$$E = m \times c^2$$

Here, then, was the all-important exchange-rate formula he had sought. He was relieved, and also gratified, because the relationship between mass and energy had turned out to be so simple, so elegant; for all its strangeness, his relativistic universe was far simpler philosophically than the old one.

Because mass and energy were interchangeable, for example, science would no longer have to deal with *two* conservation laws. Mass could be destroyed and converted into energy, and by the same token, energy could be destroyed and converted into mass. Only the grand total of all the energies and all the masses in the cosmos remained unchangeable for all time; that is, there was now only one integrated Law of Mass-Energy Conservation.

The relationship between space and time, too, had been simplified. Because of the spatial and temporal relativity in Einstein's universe, science no longer would need to distinguish between *A moving with respect to B* and *B moving with respect to A;* now only relative speeds mattered.

Furthermore, anyone fearful of having to cope with these strange new rules would not need to worry. Within the slow-moving sphere of human existence, Einstein's Theory of Special Relativity had an insignificant effect.

For example, even at speeds of *hundreds* of miles per hour, the mathematical value of Einstein's shrinking factor remained very close to 1, which meant that the various relativistic aberrations were virtually undetectable: In the realm of everyday life, there-fore, space and time and mass and energy appeared to behave nor-mally.

Even in the future, when astronauts would travel to the moon at 25,000 miles per hour, the discrepancy from normal would amount to a minuscule *five parts in a trillion*. Compared to people left behind on earth, in other words, an astronaut's impression of an inch and a second was shorter by that absolutely negligible amount.

But the news was not wholly sanguine, because for all its bril-liance, the consequences of Einstein's hike up the Swiss Alps of human curiosity were fraught with unrecognized peril. His theory had redefined for all time the spatial and temporal relationships between observers in a strange new universe; but also, in just four short decades, its unassuming mass–energy equation was to change forever the political and social relationships between countries in a forbidding new world.

EPILOGUE

Ever since Albert Einstein had discovered that such a thing was possible theoretically, scientists had sought a way to transform matter into energy. Stubbornly, scientists had persisted, partly out of curiosity and partly because they believed so strongly in the veracity of Einstein's little equation. And why not? Already they had witnessed the conversion of energy into matter: An electron whirling around inside an atom smasher grew more massive as it sped up, exactly as Einstein had predicted it would.

They had persevered also because the stakes were so high; the potential source of energy was as prodigious as the material universe itself. Once they were able to tap it, scientists predicted joyously, the world would have an unlimited supply of clean, cheap power.

Until that day came, however, people continued to harness power the old-fashioned way, by *burning* something. To generate electrical power, for example, most industrialized countries incinerated wood, oil, or coal; but the process was very inefficient. A modern power plant burning a lump of high-grade coal, for example, produced enough energy to keep one light bulb shining for only about four hours.

Coal had taken millions of years to form, the result of dead plants being buried under layers of heavy rock and compressed by the inexorable movements of the earth's shifting continents. By burning a lump of coal, the solar and seismic energies that had gone into making it in the first place were released in the form of thermal energy.

Einstein's mass-energy equation promised us far greater yields, however, if only we could find a way to convert that same lump of coal *completely* into energy (with no ash being left behind). A simple calculation revealed, in fact, that such a transmogrification would produce enough energy to keep one light bulb burning for not four hours but 1,680 *billion* hours!

In the end, it was to take scientists about 297,840 hours—thirty-four years—of hard work to convert Einstein's little formula into a blinding reality. The key to their success, furthermore, first appeared very early in the century, shortly after Antoine Henri Becquerel's discovery of radioactivity.

Back then, scientists wondered how uranium and certain other elements were able to spew out their radiation with such energy. Where did the power come from? By answering that question, scientists began to suspect, they would learn the secret of how to convert matter into energy.

During the early 1930s, scientists finally found their answer. By looking under the hood of a uranium atom—that is, by using atom smashers to give them an unprecedented peek inside the subatomic world—they were able to see what an atom really looked like.

An atom, they discovered, was not like some monolithic billiard ball; it was an elegant mechanism, with several moving parts. Essentially, it consisted of a nucleus of *protons* and *neutrons,* surrounded by an outlying swarm of electrons—not unlike a hive with its restless swarm of bees.

Neutrons, as their name implied, were extremely tiny particles that behaved as if they were electrically neutral. They did not repel each other; that is, even though many neutrons were crammed together ever so tightly inside an atom's nucleus, they made no attempt whatsoever to escape.

Not so the protons; unlike neutrons, these subatomic particles were each imbued with a positive electrical charge. Scientists long ago had discovered that similar charges always repelled one another; therefore, protons naturally resisted being confined within an atom's nucleus.

The only thing that kept them incarcerated was a slightly overpowering nuclear force—a kind of invisible nuclear glue—but even that did not always work. Within large nuclei, such as uranium, there simply were too many mutually repulsive protons for the force to keep in check; in such cases, some of the protons managed to escape.

It was like trying to hug a bunch of mattress springs. Inevitably, if the bunch grew to be too large, a few of the springs would escape one's clutches and go flying away. Those high-speed escapees from the atomic nucleus, scientists concluded, were precisely what constituted radioactivity.

Following that important revelation, scientists invented ways of actually *weighing* unstable, radioactive nuclei. Those meticulous procedures, furthermore, led to a startling observation about radioactive nuclei and brought the world one step closer to the Atomic Age.

After coughing up a subatomic particle, scientists observed, a radioactive nucleus always weighed less by an amount *greater* than the mass of the escaped particle. Evidently, radioactive particles effected their escape by robbing some of the *nucleus*'s mass and transforming it into energy, an exchange entirely in keeping with Einstein's equation.

Anthropomorphically speaking, it was as if protons were like siblings whose mutual revulsion was so intense it was palpable. In that case, one could say their combined weight consisted of their individual masses *plus* the ponderousness of their pent-up tension. After a sibling fled, therefore, the remaining family weighed less by an amount equal to the runaway's mass *plus* his share of the material tension.

In a manner of speaking, therefore, scientists in the 1930s had discovered that radioactivity was a heavy, dysfunctional nucleus's way of relieving stress. Furthermore, they reasoned, if the nucleus was so large and tense as to be on the verge of a nervous breakdown, it might be a simple matter to make it fall apart completely—and to harvest the outpouring of hysterical energy that was sure to result.

At that point, invigorated by such a well-defined game plan, scientists began to turn their attention toward uranium. Mined from pitchblende, the element uranium represented the largest atom found in Nature; its nucleus consisted of ninety-two irascible protons just busting to get out.

But how did one go about shattering a nucleus? Even for one as "big" as uranium, the task seemed to require impossibly small weapons. It was a far cry from busting apart a popcorn ball, say, considering that the nucleus of a uranium atom was a mere ten quadrillionths of a centimeter across.

At first, scientists tried shooting the uranium nucleus with an electron, but the tiny bullet proved too puny for the job. They also tried shooting it with a high-speed proton, but the repulsive force of the nucleus's own protons never let it get close enough to have any effect. Finally, in 1934, scientists tried a neutron—the only other subatomic bullet known at the time—and it worked!

The neutron, being electrically neutral, was able to infiltrate the family of mutually repulsive protons and break it apart. In the process, the radioactive nucleus was able to breathe a sigh of relief, as it were, letting out a hundred billion times more energy than one could ever get from ordinary, old-fashioned combustion.

It was, after more than thirty years, a stunning confirmation of Einstein's equation. More than that, it was an achievement akin to the discovery of fire: For the first time in history, we had found a way to liberate the energy that had remained bottled up within atomic nuclei since the time of their creation, billions of years ago.

Italian physicist Enrico Fermi was the first person to split nuclei with neutrons, though he did not realize it right away. The same was true of the French couple Irene and Frederick Joliot-Curie and a pair of German physicists, Otto Hahn and Franz Strassmann: Incredibly, all these people had succeeded in splitting the uranium nucleus, but none of them realized it immediately. Such were the vagaries of their complex effort.

It wasn't until January 1939, five years after the fact, that the physicists finally understood what they had done. However, even then, though the news of their achievement was greeted with excitement and celebration throughout the *scientific* community, it did not so much as raise an eyebrow among laypersons.

Scientists had discovered a way to relieve the uranium nucleus of its natural tension by artificial means, but for most people, it was an academic matter. Their concerns were with the political tensions abroad: During the past several years, Japan, Italy, and Germany had demonstrated an intention to take over the world.

On September 1, 1939, Hitler's Nazi army invaded Poland, and immediately thereafter the world was at war. Just as quickly, moreover, the scientists who only a few months earlier had managed to rive radioactive uranium nuclei became worried: Hitler had completely stopped the export of uranium ore from Czechoslovakia, which the Nazis now occupied. From this, the scientists guessed that Hitler's brain trust might have discovered the power of Einstein's physics.

After trying themselves without success to interest the United States Navy in their recent atomic tour de force, Enrico Fermi and others decided to consult with the one scientist whose world–class stature might make the difference. It was the summer of 1939 when the group departed for New Jersey; they were going to pay a visit to Professor Albert Einstein.

Einstein had come to the United States in 1933 and had already decided to remain when, on April 7, Hitler had decreed that Jews in Germany were to be expelled from all positions of power. Before that, in 1921, Einstein had been awarded the Nobel Prize in physics, though curiously enough, not for his Theory of Special Relativity. He had received it for his part in developing Quantum

Mechanics, a theory about atomic behavior that was even more arcane than relativity. Einstein had become more famous and worldly than any other scientist in modern times. He consorted with royalty, visited with presidents, and became the darling of the mass media—hamming it up for the cameras, even agreeing to pose with Hollywood celebrities.

In 1933, he chose to accept a position at the Institute for Advanced Studies, in Princeton, New Jersey. He had been attracted by the institute's peaceful rural setting and the companionship of old friends who also were moving there to escape the grasp of Nazis overseas. Like them, he had renounced his German citizenship, though now, in the face of Hitler's demonic megalomania, Einstein had begun to wonder whether he should do more than that.

For most of his life, Einstein had been an outsider, scientifically, socially, and politically. Time and again he had referred to himself as a "stateless" person and had ended up becoming the citizen of a politically neutral country, Switzerland.

During the *First* World War, at the beginning of the century, while Germany's army had bullied its way through Europe, Einstein's mind had bullied its way through science, permanently ravaging the intellectual landscape with one new theory after another; he was able to ignore completely the violent conflicts all around him.

"Alongside [his] work the problems of daily life did not appear very important," recalled Philipp Frank, a physicist who had befriended Einstein during those war years. "Actually he found it very difficult to take them seriously."

It was only *after* that heinous war that Einstein had been forced to take seriously the Nazis' growing influence. For one thing, the Nazis had pressured universities to abandon the promulgation of his *Jewish* physics and return to the teaching and practice of *German* physics.

One of the early converts, physicist Philipp Lenard, had insisted that the practice of science "is racial, and conditioned by blood." German physics was superior because it was, in his words: "The Physics of those who have fathomed the depths of Reality, seekers after the Truth, the Physics of the very founders of Science."

Einstein had been thunderstruck by this turn of events. All his life, he had grown accustomed to living in his own world, wherever he happened to be and whatever happened to be going on around him. But these accusations—made, as in Lenard's case, by some of his most cherished colleagues—had rousted him from that introspective bubble like nothing before, not even the Luitpold Gymnasium or the Great War itself. It had been the most revolting revelation this renegade had made in his forty years: Albert Einstein the Scientist had discovered Albert Einstein the Jew.

Now, on the eve of World War II, he was tempted once again—and mostly succumbed to the temptation—to stay detached and to concentrate on his research. But his rude awakening after World War I had caused him to realize that merely *wishing* for peace was not sufficient; one had to work for it.

Einstein had become something of a peace activist, which meant that after listening carefully to the group of anxious scientists who visited him on that day in July 1939, he was left with very mixed feelings. They were, in effect, asking his help to develop an instrument of war, the very thing he hated. And yet he realized that if the Allies could beat Hitler at creating a nuclear bomb, it might be used as an instrument of peace.

In the end, on August 2, 1939, he agreed to write a letter to President Franklin Roosevelt:

> Sir: Some recent work . . . which has been communicated to me in manuscript, leads me to expect that the element uranium may be turned into a new and important source of energy in the immediate future.

In the letter, Einstein urged Roosevelt to provide money for further research, without delay. And lest the President not understand the need for urgency, he closed with this ominous warning.

. . . that Germany has actually stopped the sale of uranium from the Czechoslovakian mines . . . might perhaps be understood on the ground that the son of the German Under-Secretary of State, von Weizsäcker, is attached to the Kaiser-Wilhelm Institute in Berlin where some of the American work on uranium is now being repeated.

When President Roosevelt read the letter, he reacted as most politicians did to any new suggestion: He formed a committee to think it over. In November, the committee reported back to the President, recommending that he do exactly what the scientists had wanted him to do in the first place.

Within days, hundreds of scientists working in universities and labs all over the United States—many of them refugees from Europe—applied themselves to the awful task of giving life to the most destructive weapon humanity had ever envisaged.

It took five years, two billion dollars, and thousands of people, but on July 16, 1945, the outcome of all the effort and expense was ready to be tested. Einstein, who throughout these years had remained at the institute, working on one of his new theories, chose not to be at the test site. The device was to be detonated in the middle of the New Mexico desert, at the Alamogordo Air Base, twenty miles from the nearest habitation.

No one knew what to expect, so the scientists were cautious in their preparations. The young physicist who had directed the design and building of the device, J. Robert Oppenheimer, was holed up in a bunker ten miles away. With him were the project's other top civilians and one of the military directors, General Thomas Farrell.

Crews had worked all night in preparation for this morning's

test, and as sunlight rose above the horizon, everyone had a clear view of the detonation tower. The countdown commenced, and when it reached zero, the explosion from the device lit up the world, much as the young Einstein himself had done forty years earlier.

"The lighting effects beggared description," Farrell would write later. "The whole country was lighted by a searchlight with the intensity many times that of the midday sun. It was golden, purple, violet, grey and blue. It lighted every peak, crevasse and ridge of the nearby mountain range with a clarity and beauty that cannot be described but must be seen to be imagined."

Oppenheimer was relieved that his project had succeeded but also frightened and sobered by what he saw: "I am become Death," he intoned sotto voce, quoting from Vedic scriptures, "A destroyer of worlds." Farrell expressed similar sentiments, explaining that following the bomb's powerful air blast came "the awesome roar which warned of doomsday and made us feel that we puny things were blasphemous to dare tamper with the forces heretofore reserved to the Almighty."

When Einstein heard the news, he was heartened by the possibility that this horrifying creation now would cow the enemy into surrendering, thus bringing about peace. But three weeks later, when he and the world saw what this new bomb had done to the Japanese city of Hiroshima—and three days later to Nagasaki—Einstein himself was cowed into having second thoughts. In retrospect, he would lament, "I made one great mistake in my life—when I signed the letter to President Roosevelt recommending that atom bombs be made."

All his life, Einstein had worshipped the mind's natural inquisitiveness about the physical world. While others throughout history had fought for their right to be free or to worship in a church of their chosing, he had fought just as strenuously and stubbornly for the right to be independently curious.

During his lifelong struggle, he had come to have utter con-

tempt for the schools of his day; he wrote: "It is, in fact, nothing short of a miracle that the modern methods of instruction have not yet entirely strangled the holy curiosity of inquiry; for this delicate little plant . . . stands mainly in need of freedom; without this it goes to wrack and ruin without fail."

For the second time in his life, however, the consequences of war had jarred Einstein into making an unexpected discovery about his personal beliefs. The A-bombs dropped on Japan—which soon ended World War II—had ended his unqualified worship of human inquiry. With his own eyes, he now saw an unholy aspect to curiosity: If the delicate little plant was not nurtured with caution and compassion, he decided, then it was *we* who would go to wrack and ruin without fail.

Following the war, Einstein withdrew into his private world one final time. His having seen the light, however, did not diminish his scientific curiosity any more than his post–World War I epiphany had made him any less Jewish; to the contrary.

Following World War I, he had become an outspoken Zionist. So much so, in fact, that in 1952, following the death of Chaim Weizmann, the Israelis were to ask Einstein to become their new president, an honor he respectfully would decline.

Now, in the wake of World War II, he became the zealous champion of another cause: Einstein wanted to come up with a single theory that could explain *everything* about the physical world, a kind of scientific oracle capable of divining all answers to all questions the human mind ever could imagine. Physicists came to call it a *Unified Field Theory*.

Over the years, though his mind remained active, his body grew older and weaker. Finally, on April 18, 1955, Albert Einstein died, in the midst of his unsuccessful efforts to find all the answers. To the end, Robert Oppenheimer recalled: "There was always with him a wonderful purity at once childlike and profoundly stubborn."

Einstein's childlike curiosity always had set him apart from the pack. Though most humans were born with an unbridled curiosity, they usually grew out of it as they grew up; in that respect, Einstein had never matured fully.

In the years to come, many looked back on this extraordinary man and questioned his involvement in the creation of the atomic bomb, much as he had second-guessed himself. The debate became even more sorrowful after 1952, when American scientists tested the world's first *thermonuclear* device—precursor of the hydrogen bomb—several hundred times deadlier than the A-bombs dropped on Japan.

Inevitably, critics blamed science—physicists, in particular—for plunging humanity into an Atomic Age that now jeopardized the future of the entire planet. It had taken billions of years for life to evolve, they fretted, and yet it would take but a few minutes for science's terrifying new weapons to wipe it all out.

While these recriminations were entirely justified, critics overlooked the all-important Darwinian assertion that during the entire course of our evolution, we had retained only those traits that *enhanced* our survivability. If the theory of natural selection was true, therefore, it was entirely possible that curiosity—far from being our nemesis—could turn out to be our salvation.

That was not to imply that, along the way, people would never be hurt or killed by curiosity. Throughout recorded history, tens of thousands—perhaps millions—of innocent persons had lost their lives for being recklessly inquisitive. But if curiosity did not ultimately serve a useful purpose, then why had such an irrepressible urge come into being and persisted to this day?

Surely curiosity was not the only double-edged trait that we had acquired over the course of our evolution. There were similar dangers inherent in those other seemingly indomitable human impulses, hunger and sex. That is, people routinely became ill or died from eating spoiled foods or from having intercourse with a diseased person, yet no one ever had proposed that we ignore our hunger or libido.

In short, the need to ask questions appeared to be in our genes, along with the need to eat and to reproduce. It was even possible that curiosity was guiding us to some specific destination—whether out there among the stars or right here on earth—to some special place and time that would teach us everything we ever wanted to know about the natural world and how best to survive in it.

If so, then Albert Einstein's curiosity had managed to lead us farther along and higher up our genetically driven scavenger hunt for answers than anyone had before. Understandably, many people today have become so anxious about the dizzy heights and precarious landscape that they wish to climb back down. But if science has taught us anything during the past 2,000 years, it is this: Retreating from the earth-shaking consequences of our scientific curiosity is as implausible as time travel and, quite possibly, as undesirable as devolution.

Index